Practical Plant Failure Analysis

Analysis

A Guide to Understanding
Machinery Deterioration and
Improving Equipment Reliability
Second Edition

Practical Plant Failure Analysis

A Guide to Understanding Machinery Deterioration and Improving Equipment Reliability
Second Edition

Neville W. Sachs, P. E.

CRC Press
Taylor & Francis Group
Boca Raton London New York

CRC Press is an imprint of the
Taylor & Francis Group, an **informa** business

CRC Press
Taylor & Francis Group
6000 Broken Sound Parkway NW, Suite 300
Boca Raton, FL 33487-2742

First issued in paperback 2021

© 2020 by Taylor & Francis Group, LLC
CRC Press is an imprint of Taylor & Francis Group, an Informa business

No claim to original U.S. government works

ISBN-13: 978-1-138-32411-4 (hbk)
ISBN-13: 978-1-03-217685-7 (pbk)
DOI: 10.1201/9780429451041

Library of Congress Cataloging in Publication Data

Names: Sachs, Neville W., author.
Title: Practical plant failure analysis : a guide to understanding machinery deterioration and improving equipment reliability / by Neville W. Sachs.
Description: Second edition. l Boca Raton, FL : CRC Press/Taylor & Francis Group, 2019. l Includes bibliographical references.
Identifiers: LCCN 2019020290l ISBN 9781138324114 (hardback : acid-free paper) l ISBN 9780429451041 (ebook)
Subjects: LCSH: Machinery--Reliability. l System failures (Engineering) l Factories--Equipment and supplies--Maintenance and repair.
Classification: LCC TJ153 .S164 2019 l DDC 621.8/16--dc23
LC record available at https://lccn.loc.gov/2019020290

Visit the Taylor & Francis Web site at
http://www.taylorandfrancis.com

and the CRC Press Web site at
http://www.crcpress.com

To Eddie Sullivan and Lauren Rose Sachs

Contents

Preface

My goal in writing this book is to help people working on machinery to understand why breakdowns and failures happen, and how a person shouldn't need to be a genius to solve the physical causes of most equipment failures.

Being involved with industrial or commercial machinery maintenance is challenging enough, and being really good at it demands a knowledge of mechanical devices and how both lubrication and corrosion affect those devices. What I tried to do in the first edition of this book was to write something that would explain the details of how the operating conditions and environment can change the appearances of failures. Hopefully this edition continues and improves on that.

From my perspective, the gratifying good news is that, over the 13 years since the first edition, a substantial number of people have mentioned that the book has helped them. The frustrating news is that I still see that a tremendous number of North American corporate managers don't appear to understand how management practices, i.e., the latent roots, contribute to the problems with their plant reliability or organization productivity. Three common examples of this are:

- In maintenance organizations – *Reducing headcount by eliminating planners.* [Do those managers really believe that every mechanic can reliably keep abreast of technical changes, order parts, schedule jobs, and do a good job of maintaining their machinery?]
- In consulting organizations – *Not involving the practitioners in assisting in developing quotes and specifications for jobs, before the quote is sent to the client.* [Merging the sales personnel's skills with those of experienced field personnel ensures a more accurate and realistic analysis of the challenges. For example, there would be a better understanding of why Type 304 stainless shouldn't be specified where high chloride concentrations are present.]
- In plant engineering departments – *Not recognizing that every plant has a huge reservoir of workforce talent that could be utilized in addressing problems.* [As we look around us, much of our world has been constructed by folks who didn't graduate from college with engineering degrees, and the data we've seen on plant failures shows that the majority involve engineering errors.]

It is hard to believe that is has been 13 years since the first edition of *Practical Plant Failure Analysis.* My life has changed greatly, in that Sachs, Salvaterra & Associates, Inc. is no more, and, after a brief visit to the corporate world, I'm back to being an individual contributor and have been an absolute failure at retiring. It is nice to be able to limit my travel, but I still have problems turning down failure analysis challenges and, thankfully, life has continued to be busy, interesting, and rewarding.

One of the benefits of not traveling as much is that it has allowed more time to study the underlying physical roots of problems and, hopefully, enabled me to do a better job explaining them. In this revision, I've tried to make the technical descriptions clearer and easier to understand, and I hope that all of the typos and grammatical errors have been corrected. There has been new material added to every section, particularly a good deal added to the section on Miscellaneous Machine Elements.

I'd really like to thank a number of folks for their assistance and help in educating me, especially Darren Bittick (ATS), Dan Carroll (Eli Lilly), Bob Buck (formerly of Morton Salt), Kim Jaynes (Oil-Dri), and Joe Park and Mike White (both formerly with Novelis). Again, I am fortunate to be able to thank a very patient wife, a skilled artist, and my favorite hiking, skiing, and kayaking partner, Carol Adamec. (She's the one who has enabled me to sit in my office and study that mechanical stuff.)

Lastly, I'm sure there are some human errors in the text and would appreciate it if you would tell me about them.

Neville W. Sachs
Camillus, New York

Author

A native of northern New Jersey, **Mr. Neville W. Sachs** attended the Stevens Institute of Technology in Hoboken, New Jersey, where he received a Bachelor of Engineering degree, majoring in mechanical and chemical engineering. After a variety of manufacturing, engineering, and supervisory positions, he joined Allied Chemical (now Honeywell International). From then until the Syracuse Works closed, he was heavily involved with plant reliability as an engineer and reliability engineering department supervisor. While there, he was instrumental in developing one of the first large predictive maintenance inspection programs in the nation, served on a number of corporate technical committees, and received a patent for a device that demonstrates several of the mechanisms of fastener failures.

In early 1986, Mr. Sachs, a licensed professional engineer, joined with Philip Salvaterra to form Sachs, Salvaterra & Associates, Inc. (SS&A), a consulting "Reliability Engineering Department for Hire". After 25 years serving as the president of SS&A, the company was absorbed by Applied Technical Services of Marietta, Georgia, and in 2014 he returned to private practice.

Mr. Sachs has conducted thousands of failure analyses and taught hundreds of failure analysis seminars across North America and Europe. He is a past chairman of the Syracuse Chapter of the ASM and is an active member of the National Association of Corrosion Engineers, the American Society of Mechanical Engineers, National Society of Professional Engineers, and the Society of Tribologists and Lubrication Engineers (STLE). In addition to being certified in several areas of nondestructive testing, his formal certifications include STLE's "Certified Lubrication Specialist".

He is a frequent speaker for programs across North America and has written three textbooks: *Practical Plant Failure Analysis – a Guide to Understanding Machinery Deterioration and Improving Equipment Reliability, Failure Analysis of Gears and Bearings Made Simple,* and *Failure Analysis of Shafts and Fasteners Made Simple.* He has also contributed significant sections to three other books concerning mechanical reliability and failure analysis and has written over 70 technical articles and papers for U.S. and European magazines and journals, primarily on failure analysis and equipment reliability. Among his honors are the RMLA's "Outstanding Contribution to the Industry" award (2019).

He and his wife, Carol Adamec, a noted sculptor, hike, bike, kayak, ski, and try to keep up with 10 grandchildren. He also enjoys being an NSP ski patroller and playing with old cars.

1 An Introduction to Failure Analysis

The people we've worked with started doing in-depth failure analysis on industrial equipment in the mid-1960s. Prior to that time there weren't a lot of industrial failure analyses, and the ones that were done were just involved with trying to understand the physical causes. The efforts of those early folks were primarily linked with an interest in improving production equipment reliability and capacity in chemical plants. From their work and a manufacturing and processing viewpoint, it wasn't until the early 1970s that a realization began to develop that the true sources of industrial problems were much more complex.

What is *failure analysis*? There are probably as many definitions as people you ask the question of, but we prefer to think of it as "the process of interpreting the features of a deteriorated system or component to determine why it no longer performs the intended function". Failure analysis entails using deductive logic to find the physical and human causes of the failure, then using inductive logic to find the latent causes. From an understanding of these "failure roots", there should be a path to the changes needed to prevent the recurrence of the incident.

Some people in industry prefer not to use the term *failure analysis*, and more than once we have heard a statement such as "We don't want our maintenance improvement (or some similar) program being driven by concentrating on failures". It's easy to understand their words but impossible to understand their logic. Most of us learn from our mistakes and, in the same manner that professional athletes use when they study game videos or farmers use in analyzing soils and crop yields, failure analysis allows us to look at our weaknesses and errors, gain knowledge from them, and try to do a better job the next time.

This book is an attempt at a manual that explains how and why mechanical machinery fails and how to solve those problems. Realizing that no single text can address all failures, this book tries to explain how the basic failure mechanisms occur, the things we all do to cause machinery problems, how to recognize those things, and what to do to prevent future similar incidents. Unfortunately, there is an almost infinite number of failure symptoms and appearances and the book can't address all of them. But it should allow the careful reader to analyze and solve by far the majority of the mechanical failures that occur in the typical paper mills, chemical plants, power plants, and manufacturing facilities.

THE CAUSES OF FAILURES

Why are there premature equipment failures? When the people closely involved with the failure are asked this question, they almost always say it is "the other guy's fault". If one were to ask a plant millwright or a maintenance mechanic, the most likely answer to that question would be "operator error". But if the same question were asked of an operator who worked in the plant with that millwright, their answer might be "because it wasn't properly repaired". At times there is some validity to both of these answers, but the honest and complete answer is always much more complex.

It would be nice and neat if there were only one cause per failure, because eliminating the problem would be easy, but in reality, there are multiple causes to every equipment problem. Unfortunately, there are many people who believe that there is only one cause for a failure. However, look at the analysis of any well-studied major disaster and ask if there was only one cause. Was there a single cause for the BP oil well disaster? ... Three Mile Island? ... the Exxon Valdez mess? ... Bhopal? ... Chernobyl? ... a major airplane crash? The analyses of these and other, well-recognized and extensively studied failures show that they all have multiple causes. Then, why would any intelligent person believe a typical pump or fan failure would be different? In the case of Three Mile Island, there were three huge studies, each commissioned by one of the responsible groups. All three of the studies said there were numerous causes but that it was "primarily the fault of the other two organizations". In doing failure analyses, it is often amusing to listen to the management staff talk about how the workforce employees "messed up" *without any recognition at all* of how their engineering and management practices were involved.

At an international conference on failure analysis, a presentation was made on the causes of aircraft equipment failures. The presentation data showed:

- 30% – Manufacturing Errors
- 26% – Design Errors
- 23% – Maintenance Errors
- 18% – Material Selection
- 3% – Operation

During the question-and-answer session after the presentation, a member of the audience asked the speaker why they had listed only one cause for each failure when there were usually multiple causes. The speaker agreed with the questioner's point, but then said, "There was only one blank on the form". This answer is a quote and an interesting testimony to the general public's lack of perception.

When people discuss cases that have been carefully studied, such as those listed earlier, they almost always agree that there are multiple causes for each. Yet when directly involved with a failure, the ability to be objective seems to disappear and, ignoring reality, many people come up with conclusions such as those mentioned in the presentation above. They then take this data, draw an attractive pie chart or bar graph, and point to a nice neat single cause for every failure ... when an honest analysis clearly states that is neither true nor logical.

Two comments:

A. At a later time in the session, another group analyzed the same basic airplane equipment failure data set but reached very different conclusions. They too sorted the data with the idea of a single cause for each failure!

B. One of the questions that I find interesting is, "Why don't these people recognize that many failures have more than one physical cause and all failures have more than one human cause?" I suspect that there is an innate human desire to neatly categorize problems and answers, but the answer to this lies with people more skilled than I.

Trevor Kletz has written a fascinating series of books detailing events leading to many of the world's major industrial disasters.*† In his approach to failures, he uses a three-tiered system for classifying the causes, and we use a similar method. In our approach, there are three general classes of failure causes or roots, and, until the investigator understands each of these classes and their inevitable interaction, they do not truly understand how and why the failure occurred and they won't have the ability to prevent another.

The divisions that we have used for these roots (causes) and some examples are:

- *Physical root* – This is the physical mechanism (or mechanisms) that caused the failure, i.e., the shaft failed from rotating bending fatigue complicated by both corrosion and a significant stress concentration.
- *Human root* – This is the "inappropriate human intervention" that resulted in the physical roots. Continuing the example above:
  1. When the shaft was designed, the engineer didn't expect the corrosive conditions and used an alloy where corrosion rapidly and significantly reduced the fatigue strength.
  2. The machinist who made the shaft cut a sharp corner where there should have been a gentle radius.
  3. The millwright who installed the machine misaligned it, causing vibration and an unanticipated bending load on the shaft.
- *Latent root (system weakness)* – These are the corporate policies or actions that allow the "inappropriate human action". Again, pursuing the example above:
  1. The engineer was pressured by supervision to rapidly complete the assignment and didn't carefully analyze the full range of conditions that the machine was going to experience. In addition, he was only two years out of college as a mechanical engineer, didn't do well in his materials classes or really understand the use of different alloys, and his supervision had no idea of these weaknesses.

---

* Kletz, Trevor, *Learning from Accidents* (3rd edition), Taylor & Francis, 2001, ISBN-10: 075064883X.
† Kletz, Trevor, *Lessons from Disasters*, Gulf Professional Publishing, 1993, ISBN-10: 0884151549.

2. After the machinist made the shaft, it was shipped to the plant for instal-
    lation. Unfortunately, at the plant there was no receiving inspection for
    materials used in maintenance and the shaft was accepted with the error.
3. The maintenance management had stated that the millwrights should be
    using laser alignment equipment, but the policy wasn't enforced, and this
    mechanic hadn't been to the alignment class. (It was a "hot job" and had
    to be back in operation ASAP, so they couldn't waste time on alignment.)

One truly significant industry dilemma is that problems that are not the result
of catastrophic failures are frequently not recognized at all! Often normal corrosion and wear, i.e., problems where slow wastage destroys a component or a
piece of equipment, are not recognized as opportunities for improvement and are
ignored. An unexpected leak in a pipe or vessel may be recognized as a failure
but the long-term deterioration of a tank or the replacement of a set of V-belts
every 12 months is thought of as being "just part of the cost of doing business".

Much of the routine maintenance a plant or facility falls into this "ignored
failure" category. During an assignment in a plant that was generally accepted
in their industry as having an excellent maintenance program, one of their
instrument technicians found that they were spending over $300 per day
replacing gas system filters that should have cost less than one-tenth of that.
In a similar plant, extensive interviews with maintenance personnel found that
almost 20% of the routine maintenance budget resulted from recurring items
that should never have happened in the first place. In another plant, 40% of the
maintenance costs involved piping leaks. In all of these situations, the problems had been in existence for years but had never been recognized.

ROOT CAUSE ANALYSIS (RCA) AND UNDERSTANDING THE ROOTS

The concept of root cause analysis (RCA) has many interpretations. We use it to
understand the multiple physical, human, and latent causes of a failure, but many
folks like to stop at the physical causes. As you can see from the three tiers of roots
listed above, if the problem is solved for just the physical causes of the failure, all
that has been done is to determine why that one incident occurred. Going into greater
depth and finding the human roots allows one to change the human behavior and
eliminate a group of failures. But discovering and correcting the latent weaknesses,
the way the plant or corporation is operated, results in whole classes of failures being
eliminated. As examples of this, look at the changes in automotive reliability since
the introduction of Japanese imports in the 1970s or the improvements in computer
reliability and cell phone communications in the last 15 years.

PHYSICAL ROOTS

The physical root is the mechanism (or mechanisms) that caused the part to fail,
and it may be fatigue, overload, wear, corrosion, or any combination. The importance of understanding the physical root or roots cannot be overstated because, if the

analysis doesn't start with accurately determining the physical roots, how can the actual human and latent roots be detected and corrected?

Most of this book is about the physical roots of failures, and there are times when the cause is a single physical root; however, our data from years of analysis shows there is frequently more than one physical mechanism involved. In a relatively limited review of our detailed field failure analyses, we found an average of about 1.4 physical roots/failure.

HUMAN ROOTS

The human roots are those human errors that result in the mechanisms that caused the physical failures. A non-industrial example of a human root of a problem would be an automobile driver's use of a cell phone and the effect on accident rates. Three studies on cell phone usage show:

- A 1997 article in the *New England Journal of Medicine* states that motorists were four times more likely to be involved in an accident when using cell phones.*
- A 1999 report by the Center for Urban Transportation analyzed a number of studies and found that mobile phone use while driving increased the chance of an accident anywhere from 34% and to more than 300%.†
- A 2009 report by the Virginia Tech Transportation Institute found truckers using text messaging with a mobile phone had 23 times as many accidents as their non-texting peers.

Other studies have reported similar data. It seems fairly obvious that these drivers don't intentionally have collisions, and that it is the effect of cell phone use and the resultant distraction that results in both an increase in the rate of human errors and the increased accident rates.

Statistics on the frequency of human error in industry are difficult to find; however, an article in *Chemical Engineering* magazine stated the following incidences for a variety of human errors in industry.

1. Industrial Activities
   - Critical routine task – 1/1000
   - Non-critical routine task – 3/1000
   - General error rate for high stress rapid activities – 1/4
   - Non-routine operations (startup, maintenance, etc.) – 1/100
   - Checklist inspection – 1/10
2. General Human Error
   - Of observance – 1/50
   - Of omission – 1/100

---

* Redelmeier, D. A., Tibshirani, R. J., "Association between Cellular-Telephone Calls and Motor Vehicle Collisions", *N Engl J Med.* 1997, 336(7):453–458.
† *USA Today*, October 19, 2000 (from a 1999 report by The Center for Urban Transportation).

There are several texts on human error, and the data shown below is by Swain and Guttman and provides an estimate of the frequency of human errors in both plant and everyday situations.* Parenthetical values indicate 5th and 95th percentiles.

- Select wrong control in a group of labeled identical controls – 0.003 (0.001 to 0.01)
- Turn control in wrong direction in a high stress situation where design is inconsistent – 0.5
- (0.1 to 0.9)
- Operate valves in correct sequence [less than 10 sequences] – 0.01
- (0.001 to 0.05)
- Failure to recognize an incorrect status when checking an item right in front of you – 0.01 (0.005 to 0.05)

The number of errors cited in Swain and Guttman's text is daunting; however, Charles Latino,[†] the late president of the Reliability Center, Inc., Hopewell, Virginia, and a noted human reliability expert, frequently stated, "Human error experts say that the average person makes about six significant errors per week". (With regard to this, a *significant error* is defined as one where there could be a meaningful economic loss or personal injury requiring first aid treatment. Also, there have been numerous studies of human error, some of which indicate that Mr. Latino's figures may be conservative.)[‡]

In analyzing for the probability of human error, there are many conditions that affect the individual's performance. Some of these are the number of times that the task has previously been completed, the complexity of the task, time pressures on the individual, the working environment, etc. In addition, each of these conditions has modifiers. For example, the first few times that a person attempts a complex task, the individual could feel challenged by the assignment and would pay very careful attention to every detail, but later on, however, after a great many repetitions, they might be less attentive and miss critical indications.

Trying to understand the reasons for these errors is an almost endless task. However, one significant point is that we, as humans, have an unrealistic opinion of our abilities and invariably overstate them. In Tom Peters' and Robert H. Waterman, Jr.'s work *In Search of Excellence*,[§] he describes surveys where students are asked to evaluate their abilities. The self-evaluations on "the ability to get along with others" and "leadership ability" seem grossly overestimated, but we don't have a way to comparatively evaluate these skills. The students were all industrial and business majors and, when 94% of these students rate themselves as having above-average athletic ability, and 60% rate their athletic ability as being in the top 25%, we know they (and, unfortunately, the rest of us) don't have a realistic assessment of their skills.

---

* Swain, A. D. and Guttman, H. E. *Handbook on Human Error Reliability Analysis*, NOREG/CR-1278, SAND80-0200, Sandia National Laboratories.
† Latino, Robert J. and Latino, Kenneth C., *Root Cause Analysis*, CRC Press, 1999, ISBN: 0-8493-0773-2.
‡ http://panko.shidler.hawaii.edu/HumanErr has an interesting and extensive list of common human error frequencies for a number of areas.
§ Peters, Tom and Waterman, Robert H. *In Search of Excellence*, Grand Central Publishing, 1988, ISBN-13: 9780446385077.

In an effort to understand more about this persistent tendency for us to overestimate our capabilities, for 21 years as part of the introduction to my seminars, I conducted a series of surveys of industrial and commercial workers asking them to evaluate their capabilities. The survey questions are:

1. We all know that most accidents and failures result from multiple causes. These multiple "roots" act in a chain that ends up in an "undesirable event". How often do you think you make an error that could contribute to this chain and cause a significant economic loss or injury? (Significant is either requiring medical treatment or painful to your finances.)

More than once per day	_____	Once per day	_____
Once per week	_____	Once every two weeks	_____
Once per month	_____	Once every two months	_____
Once every three months	_____	About every six months	_____
Once per year or less	_____		

2. How often does the average person you know make a significant error?

More than once per day	_____	Once per day	_____
Once per week	_____	Once every two weeks	_____
Once per month	_____	Once every two months	_____
Once every three months	_____	About every six months	_____
Once per year or less	_____		

3. Compared to the other people you work with, i.e., people in jobs generally similar to yours, how do you rate your job skills? (If you are a mechanic and compared yourself with other mechanics, where would your skill level be?) Please circle the number that best describes your skill level.

0%   10%   20%   30%   40%   50%   60%   70%   80%   90%   100%
(lowest)   (highest)

4. Compared to the *other people you work with in your plant*, where would you rate your safety awareness?

0%   10%   20%   30%   40%   50%   60%   70%   80%   90%   100%
(lowest)   (highest)

The result of the survey from well over 3000 seminar attendees found the following:

- Compared with Latino's statement that the average person made six significant errors per week, the attendees felt they made a significant error about once every five months.
- The average worker indicated their skills were at about the 72nd percentile level, and their peers made about twice as many errors as they did.

- The average worker felt their safety awareness was at the 83rd percentile.
- In one particularly hazardous occupational group, where almost *10% of the employees had a lost time injury every year*, the average worker rated their safety awareness as being in the 83rd percentile.
- In 21 years of using this quiz, fewer than 45 people described their skill level as being below average.
- When we conducted this survey with engineers and engineering managers, the average answer to the first question was that they thought they made a significant error about once every 100 days, off from reality by only about a factor of 100!
- *No one* ever said they were below average in safety awareness!

Consistent with thes studies above, one of the things that we have found in conducting failure analyses is that the typical (and frequently unstated) reaction to a failure is that "It must be the other guy's fault". Maintenance people *really believe* the problem is operator error. Operators *really believe* the maintenance people don't fix the machines properly, and it is not uncommon for engineers and supervisors to believe everybody else is at fault. In an effort to better understand more about the causes of equipment failures, we reviewed 131 detailed failure analyses that our company conducted during a three-year period.

From the 131 analyses, the *major* physical failure mechanisms were:

23 Corrosion	18%
57 Fatigue	44%
15 Wear	11%
17 Corrosion fatigue	13%
19 Overload	15%

(Percentages do not equal 100% due to rounding.)

Two important notes:

1. In defining these five categories, there is the possibility of confusion between corrosion fatigue and fatigue. Our practice was to assign fatigue as the primary mechanism in those cases where the component would have eventually failed and corrosion was not needed to affect the failure. In those situations where the component would not have failed without the action of the corrosion, i.e., there was cyclic loading but it was not severe enough to cause cracking without corrosion, the cause was listed as corrosion fatigue.
2. The list shows only the primary failure mechanisms and many of them actually had multiple physical roots.

THE HUMAN ERROR STUDY

Those 131 failure analyses all involved plant site visits and interviews with multiple plant personnel. In the study we realized that, because of our various clients' directives, we couldn't guarantee that all of the human roots had surfaced, but we are reasonably confident that a good representation was found.

The study divided the major human errors into six categories as follows:

- *Design* – There are two types of design errors, those of omission and those of commission.
  1. An omission error is one where the engineer/designer fails to notice a design flaw that results in a premature failure, such as a stress concentration or an improper material selection.
  2. A commission error is one where the decision is made to use an otherwise acceptable design under conditions that lead to failure. An example of this could be a situation where a machine that is operating well at 500 units per day without a serious study is increased to an operating rate of 550 tpd and, as a result of the increase, failures occur.
- *Manufacturing* – Where a machine that is properly specified and designed is improperly manufactured.
- *Maintenance* – Maintenance errors occur when a well-designed and manufactured machine is improperly repaired or the repaired unit is improperly reinstalled. An example of this is when a millwright decides to slip fit, instead of shrink fitting, a coupling that then enables torsional fatigue forces to crack the shaft.
- *Installation* – When a well-designed and manufactured machine is improperly installed.
- *Operating errors* – Occur when a machine is run at conditions outside normally accepted ranges.
- *Situation blindness* – The proper word for this is "nescience", but this category could also be called "Can't see the woods for the trees". It pertains to those errors that occur when an obvious problem situation is ignored for an extended period of time, culminating in a major failure. (This category deserves a special recognition because it is something that we all tend to think the "other guy" is responsible for, and we rarely recognize it in ourselves. Yet there are times when all of us are blind to answers that are obvious to the other 99.999% of society. We have all seen times when another driver stops at a green light. And we've all seen drivers accidentally drive through red lights. But, fortunately, we've never done either one of these actions?)

In developing these categories, the first five were relatively straightforward. However, after a review of what had actually happened, we felt that an additional category was warranted. We use the term "Situation Blindness" to describe those failures where the gross neglect of a well-recognized but relatively minor problem allowed it to grow into an expensive and catastrophic failure. An unfortunate disaster, but a good example of management's blindness to their weaknesses, was a 1500 kW (2000 hp) reducer failure in a large paper mill. It was in a plant where procedures said that every operator was responsible for checking the lubricant level on every machine in their area on every shift. The reducer was located on a pedestal that was remote from the operator's control room and awkward to get to: down a flight of stairs, across the operating floor, and ten feet up a ladder. The reducer had been leaking heavily for 12 months, and both the area around it and the sides of the pedestal were saturated with oil. Certainly, the direct causes were the leaky seal and the lack of operator attention (failure to replenish the oil), but we felt there also had to be recognition of the general lack of insight on the part of the plant operating and supervisory personnel. When a critical process depends on human reliability, there will be failures.

Another example involves the repeated failure of a 225 kW (300 horsepower) crusher. When we were called in to do the failure analysis, there were many complaints about the poor design of the machine and lack of support from the vendor. Yet after a couple of days of detailed analysis and long conference calls with both the manufacturer and the plant, we found that the plant had ignored the crusher's poor performance in an earlier installation, had a local shop repeatedly repair the machine in a manner the vendor didn't agree with, had installed the machine incorrectly, had tried to use the machine for an application it wasn't designed for, and the vendor had recommended a much larger machine for the application.

The study found a total of 276 major human errors and hundreds of minor contributory causes. The values below indicate the number of times that this specific type of major failure cause occurred. Note that there is almost always more than one contributory cause, and a typical example of this is a calciner shell failure that was caused by pitting corrosion. The pitting resulted from the combined effects of a welding error during fabrication and an operating error.

The total number in each category shows:

93 Maintenance errors
90 Design errors
34 Operational errors
28 Manufacturing errors
20 Original installation errors
11 Situation Blindness errors

Recognizing that the distribution of these major contributors wasn't uniform, we then looked at the actual number of failures affected by each category, i.e., we sorted to see how many involved design errors, maintenance errors, and so on. In assessing the major failure causes, the totals below indicate the percentage that the specific groups of errors affected:

59% – Design errors
38% – Maintenance errors
24% – Operating errors
16% – Original Installation errors
12% – Manufacturing errors
9% – Situation Blindness errors

We began the project with some suspicions as to what we would find, but the results of the data analysis were not as expected. To ensure that there were minimal errors in the study, both the analyses and the data summaries were reviewed several times. Looking at the year-to-year summaries there were variations, but they were all within the same general range and we've also found others with similar results.

During this same period, we were also deeply involved with several true *root cause analyses*. These indicated that when a major industrial failure is taken to the point where all the possible causes are examined, there were usually somewhere between 8 and 12 significant roots. The distribution of the physical and human roots tends to fall along the same lines of what we saw in the root cause investigations (RCIs) with the exception that there are additional minor roots discovered. The greatest difference is in the discovery of the multiple latent roots.

LATENT ROOTS

These are the systems or the practices that allow the human and physical roots to exist. For example, most companies make tremendous efforts to assure good product quality. The inspection criteria for both incoming and outgoing materials are very strict and one defect in a million production units is frequently not acceptable. Yet very few of these same companies have receiving inspection programs on their maintenance supplies, and studies that we've been involved with show that typically 4% of what is received on the maintenance dock is not what was on the purchase order. Understanding latent roots can be difficult, especially if you are in the midst of them, so a medical example may yield a better viewpoint of how these latent roots effect failures.

If you had a heart problem, what type of doctor would you consult? Why wouldn't you go to a dermatologist? The dermatologist has the same pre-med education as the cardiologist and then they go through the same four years of medical school. But ... there is a little difference in residency and specialization and *when it comes to YOUR health you want to be sure the doctor is correct*.

Nevertheless, in many plants we see a management philosophy, where the directive has been to reduce apparent costs, that effectively states "Engineers are engineers" and fails to recognize the differences between chemical engineers, mechanical engineers, mining engineers, etc. Certainly, a chemical engineer can get out their handbook and design a mechanical drive, but what would your reaction be if your doctor practiced out of a handbook? This type of management thinking denies the value of experience and is the consequence of having systems that don't accurately assess the cost of these failures. As a result, they have more design errors and more reliability problems in both their production operations and their product.

Some simple examples of management lack of awareness, each of which cost more than US$175,000 (in 2017) concerning engineers are:

1. The chemical engineer who modified the design of a conveyor drive and didn't realize the position of the drive pinion had an effect on the forces on the conveyor shaft.
2. The mining engineer who suffered through repeated fastener failures without understanding how fatigue loading affects bolts.

3. The mechanical engineer for a chemical treating facility who didn't understand intergranular corrosion.
4. The design engineer who specified a reducer lubricant that was incompatible with the grease that came in the supplied ball bearings.

(These latent errors are everywhere and without them life would be much less challenging. Look at the situation in Florida in the 2000 election when George W. Bush and Al Gore were contesting the U.S. presidency. Initially in Palm Beach County almost 20,000 ballots were thrown out because of inaccuracies, about 5% of the total ballots in the county. When first asked about this, the response by a local elections representative was that they typically have about 3% of the ballots thrown out and this small increase was nothing unusual. What are the fundamental thought processes of a person who says that a 66% increase in the number of defects isn't unusual, or of the people administering an election procedure that repeatedly ignores 3% of the voters?)

THE MULTIPLE ROOTS AND HOW THEY INTERACT

In many years of doing failure analyses across North America, we have seen that there is a management tendency to ignore the latent roots and blame the individual. The result of doing this is that it doesn't solve the problem and doesn't stop it from happening again. There is solid data that shows that the typical plant failure has four to seven physical and human roots. If a failure has six roots and action is taken to correct only one of them, i.e., censuring or penalizing the person involved, what happens to the sources of the other five roots? They come back to haunt and create another failure.

WHY MULTIPLE ROOTS ARE FREQUENTLY MISSED

Early in my career as a Reliability Engineer I was told, "One of the real problems with machinery reliability is that it isn't what we know we don't know that hurts us. What is really painful and expensive are those things we think we know that we really don't understand".

Looking at the fact that there are multiple causes to all failures, Charles Latino often spoke of what he called the *Error-Change Phenomena*. At this point, we all know that the typical plant failure is the culmination of a sequence of events. If a person recognizes that this sequence has the possibility of causing a failure and makes a change that breaks the sequence, that failure won't happen. However, when the chain is broken by random incident, then the probability of the failure recurring is significant (Figure 1.1).

In one example of this, a process pipe in a huge Canadian chemical plant broke, resulting in a spill into an adjoining stream that caused an extensive and expensive fish kill. In the analysis of the failure, we found that they had had two similar failures in the past ten years. After each of the previous failures the equipment had been repaired, yet the failure recurred. As in most plants, the normal practice had been for the plant people to look at the failure, repair it as fast as possible, and get back into production. After the latest failure, a detailed investigation of the incident using

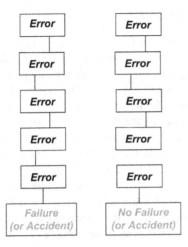

FIGURE 1.1   The Error-Change phenomenon with the first chain showing the complete failure sequence. In the second sequence, the error chain is recognized and an appropriate intervention has broken the chain.

a logic tree found that there were nine major physical and human roots, and it was only after the analysis that the plant personnel recognized that some of the causes had been there for many years while others were relatively recent. They had a chain of nine events that caused the failure and any time the nine occurred in the correct sequence the failure returned. Only by eliminating several of the roots could the problem be permanently eliminated.

The incident involved the failure of the piping below the storage tank. In the latest event the pipe cracked between the tank and the discharge control valve and the flow couldn't be shut off. Among the major human roots that the analysis found were: the pumps were run until they cavitated and were sometimes allowed to cavitate for hours; the piping system had a resonance that was excited by the pump vibration; the pipe repair procedures resulted in piping that was inherently much weaker than the original design; the pumps were being improperly repaired; and the bases were corroded and greatly weakened. ... If any one of these were corrected the system would be back in operation ... and waiting until the next time to fail.

One looks at this failure and says, "How can the plant operators allow severely cavitating pumps to run for hours. Don't they understand the damage that's being done?" Yet less than two months later we were 1500 miles away, investigating a similar failure in another country. Again, the operators allowed the pumps to cavitate for hours and the pumps were destroyed. In both sites the supervising people were fairly intelligent but had never been taught that severe starvation cavitation can destroy a set of pump bearings in less than a day. Two excellent examples of both the latent roots and the value of education.

THE BENEFITS AND SAVINGS

The cost savings that result from failure analysis can be compared favorably with the benefits of any other effective system analysis. Essentially, we are looking at a problem, understanding in complete detail how and why it occurred, and then trying to prevent it from occurring again. The results should be increased asset availability, increased throughput, and reduced costs; however, the magnitude of the benefit is directly a result of the amount of site support. We have been associated with several failure analysis/reliability programs that were *only* supported by the maintenance organization. In those programs the carefully documented annual returns were in the range of six to ten times the program cost. However, when supported *across the organization*, other programs have shown production outputs that have steadily increased over the years with no capital investment and reduced maintenance and operating costs. In an active and well-supported reliability/failure analysis program, it is not surprising to see annual returns of more than 50 times the cost.

(The plant where I learned about *The Reliability Approach*™ was over 90 years old with a mature process that had not been changed for years. The sales volume of the plant was about half a billion dollars per year and it had historically lost between 17 and 18% of the production capacity due to downtime. During the first six years of this reliability program, despite what was essentially no money was spent on capital equipment and there was a reduction in maintenance staffing and expenditures, we were able to increase effective production time by about 1% per year, and the actual output increased by more than that. The net was about a one-third reduction in downtime for a process that hadn't materially changed in 40 years. Figure 1.2 shows the actual results of the program.)

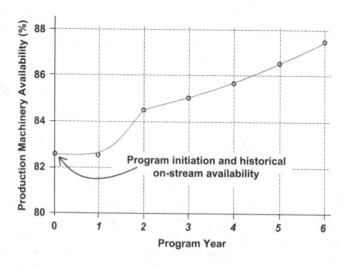

FIGURE 1.2   The effect of a formal reliability/failure analysis program. (After this, availability continued to increase but there were site changes that make later data inaccurate.)

Industry personnel with extensive experience in improving plant reliability generally think of dividing analyses into three categories in order of complexity and depth of investigation, and they are:

1. *Component failure analysis (CFA)*, which looks at the piece of the machine that failed, e.g., a bearing or a gear, and determines that it resulted from a specific cause such as fatigue or overload or corrosion and that there were these x, y, and z influences. This type of analysis is solely designed to find the physical causes of that failure and the cost is typically in the range of $2,500 to $8,000 (2018 US$).

2. *Root cause investigation (RCI)* is conducted in greater depth than the CFA and goes substantially beyond the physical roots of a problem to find the human errors involved. It stops at the major human causes and doesn't involve management system deficiencies. RCIs are generally confined to a single operating unit, i.e., a plant or a large department, and the total cost is in the range of $10,000 to $35,000.

3. *Root cause analysis (RCA)* includes everything that the RCI covers plus the minor human error causes and, more importantly, the management system problems that allow the human errors and other system weaknesses to exist. It is not uncommon for an RCA to extend to plant sites other than the one involved in the original problem, and the total cost could exceed $100,000.

Although the cost increases as the analyses become more complex, the benefit of the RCA is that there is a much more complete recognition of the true origins of the problem.

Using a CFA to solve the causes of a component failure answers why that specific part or machine failed and can be used to prevent similar future failures. Progressing to an RCI, we find that the cost could be four or five times that of a CFA, but the RCI adds a detailed understanding of the human errors contributing to the breakdown and can be used to eliminate groups of similar problems in the future. However, conducting an RCA may cost well into six figures and require several months. These costs may be intimidating to some, but the benefits obtained from correcting the major roots will eliminate huge classes of problems. The return will be many times the expenditure and will start to be realized within a few months of the formal program implementation.

One thing that has to be recognized is that because of the time, manpower, and costs involved, it is essentially impossible to conduct an RCA on every failure. The cost and possible benefits have to be recognized and judgments made in order to decide on the appropriate type of analysis.

A confusing point about RCA is that some industry personnel consider a component failure analysis as an RCA and some people consider a root cause investigation as an RCA. In our opinion, if a person wants to really know the reasons why a problem occurred, they shouldn't stop looking at some arbitrary point, then say they know all the causes. Certainly, there will be times when your position or the plant management places a limit on the extent of an investigation. In those cases, the analysis should be accurately described as what it is and not called a root cause analysis.

It is impossible to write something from a distance in a book and say your savings will be $xx; however, we can say that we've never seen a formal failure analysis program where the annual savings have not been significantly greater than the annual program cost. Some of the companies with which we've worked have claimed program savings that initially seem outlandish, but consider Figure 1.3. The figure is a graph of production vs. time for a large manufacturing plant. As you can see, there are the daily swings in output while the plant is averaging 2200 units per day. Every year there is a stretch when the plant makes 2700 units/day for several days in a row, so we know the *true production capacity is at least 2700 units/day.*

Then one day there is a major failure, a disaster when production ceases. Everybody pitches in and works all sorts of hours until production is restored and then the failure analysis begins. This *sporadic failure* is studied in great detail, sometimes with outside specialists, until even the janitors know how much it cost. In the meantime, production has gone back to its daily swings, 1600 units today, 2700 tomorrow. What management fails to recognize is that the annoying small day-to-day failures are costing them over 400 units per day, the difference between their average and their realistic capacity.

These day-to-day annoyances that cause the rate to swing back and forth are usually called *chronic failures.* They are rarely studied, except by companies truly committed to reliability, and are the source of amazing production increases and cost savings. Although the numbers above are from one specific plant where *The Reliability Approach*™ was implemented, we have worked at several others where

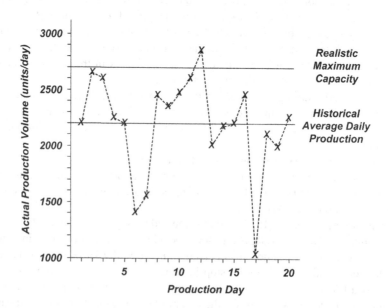

FIGURE 1.3   A plant's actual production history.

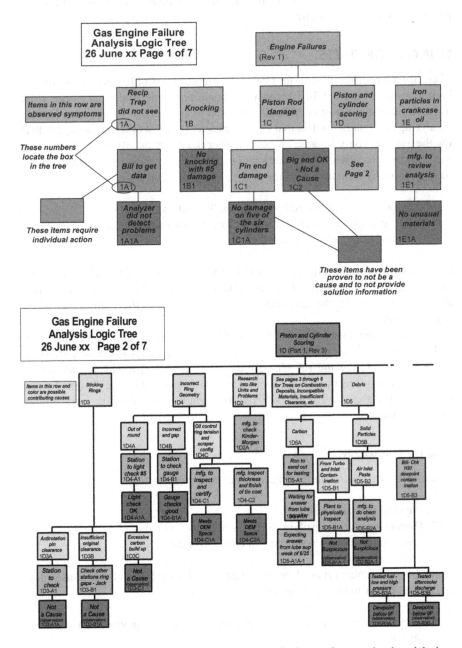

FIGURE 1.4   Here we show two pages of a seven-page logic tree that was developed during a very complex investigation into the causes of some large engine failures. (We should note that the major cause was eventually traced back to incorrect engineering data from a major supplier.)

the numbers are essentially identical and benefits of more than 20 to 50 times the program cost have been realized.

USING LOGIC TREES

There are many different tools for finding the various roots of failures. These include logic trees (see Figure 1.4), Ishikawa (fishbone) diagrams, and other systems that use an infinite number of combinations as well as a variety of copyrighted formats. Analyzing all of them, we feel that logic trees offer the most efficient method of finding the roots, not only because of their superior flexibility but also because the strict use of the tree requires the verification of each step. This effectively counteracts the most noteworthy weakness of many of the alternatives that allow:

- A hypothesis to be stated but not formally confirmed.
- Opinions to be stated concerning a possible cause without carefully considering the other possible alternatives.

Using logic trees correctly requires both the verification of the hypothesis and the consideration of all possible contributing factors. The benefit is a measurable improvement in the accuracy of the analyses and the confidence of the organization.

Logic trees are a very effective tool for investigating the human and latent roots and, in those applications, we tend to use them in the classical manner. However, when it comes to solving the physical causes of a failure, modifying the tree to involve the top box as the failed component and the next row as the observed symptoms is very effective. The investigative procedure is explained in much greater detail in Chapter 2.

Chapter 1 Summary

In summary, we see that failures have three levels of roots:

1. Physical mechanisms
2. Human errors that result in the physical causes
3. Latent roots that allow the human errors to occur

A number of different types of analyses fall under the umbrella of root cause analysis, and we prefer to look at them as:

- CFA – A component failure analysis that uncovers the physical reasons for failures.
- RCI – A root cause investigation goes deeper than a CFA and finds the major human roots but doesn't find all the human roots and doesn't look at the latent roots.
- RCA – A root cause analysis finds all of the contributors to a failure and, although the investigation is costly, typically pays for itself many times over.

Human error is unavoidable, and we all make errors that contribute to failures and accidents.

A serious difficulty that failure analysts frequently experience is a management limit as to the depth of the analysis and it is often very difficult to affect the latent roots.

BIBLIOGRAPHY

The Baby Book, Reliability Center Inc., Hopewell, VA.

2 Some General Comments on Failure Analysis

THE FAILURE MECHANISMS – HOW THEY OCCUR AND THEIR APPEARANCES

As shown in Figure 2.1, there are only four general categories of failure mechanisms: overload, fatigue, wear, and corrosion, and all mechanical equipment failures can be fitted into these categories. Sometimes there is confusion because some professionals say that a specific mechanism is in this category while others insist it is in that grouping, and sometimes more than one mechanism is involved, but the simple truth is that there are not a lot of different failure mechanisms. The following pages will try to take a practical approach and explain the basics of not only how these mechanisms function but also how they can be recognized.

Looking at the difference between the right and left sides of Figure 2.1, it is apparent that the corrosion and wear mechanisms, with a very few exceptions, remove material from the parent piece until it fails from the normally expected stresses. On the left side of the chart, the mechanisms essentially overstress the material until it fails. However, one caution is that experience teaches that in many failures, there is more than one mechanism at work.

WHEN SHOULD A FAILURE ANALYSIS BE CONDUCTED AND HOW DEEPLY SHOULD IT GO?

It really depends on the possible consequences. When a situation develops where there is the possibility of a serious injury or a serious environmental event, a full root cause analysis (RCA) is always warranted. The rest of the time the decision should be based on the costs and the possible implications. (Some of the most valuable RCAs in industry have been conducted on near misses!) The situation has to be analyzed to determine both the failure cost and how long the part should have lasted, then analyzed to see if the cost of conducting the failure analysis and improving the installation are worth it. Then, depending on the potential, a decision to conduct either a root cause investigation (RCI) or a component failure analysis (CFA) should be made.

Failure analysis is time consuming and, to maximize the return on the effort, the analyst should start out with an idea of how long the machinery is designed to last and the possible return on the time invested. For example, in one plant that had 2200 motor-driven pieces of equipment the management decided to initiate a program of

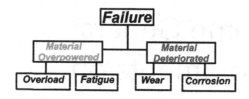

FIGURE 2.1　The four basic failure mechanisms.

motor failure analysis. At the start of the program, the average motor lasted just over 5.4 years, and eight years later the average motor life was 10.9 years. Looking at the cost benefit from the project, in 2017 U.S.$, assume that the average motor was 25 hp, cost $1800, and the manpower/record keeping/installation/etc. cost was also $1800. The annual savings from reduced motor replacements was over $700,000 – every year, and this doesn't consider the production increases, etc. that resulted from the smoother operation. From this, one can see that this project had great benefits and the key to successful failure analysis is that the benefits should greatly outweigh the costs.

How Long Should It Last?

An important key to starting a program is realizing how long many of the common industrial components should last before major maintenance or replacement is needed and Table 2.1 shows the typical lives, running 24 hours/day and 350 days/year, that should be expected with normal design loads.

Many years ago, a wise manager said to me, "If you get really good at something in maintenance, you're doing it much too often and it's time for a failure analysis". The undeniable logic behind his statement is that most parts are designed to last many years and should not be changed on a short-term schedule. But, in the heat of battle, it is commonplace that humans put up with frequently changed parts, and we often hear of personnel who are proud of their ability to rapidly replace failed components but don't understand why the failures occur.

Before starting a failure analysis, it should be recognized that this "typical expected life" is not a constant. Economic, operational, and social pressures have had a huge effect. For example, in the 1960s it was common to specify gear reducers that lasted ten to 12 years without requiring maintenance while many of the recent applications are about one-third of that. On the other hand, typical automobile engine life in the 1960s was around 160,000 km (100,000 miles), and today it is well over 320,000 km (200,000 mi.).

Whatever the component is, recognize that the design life is not a hard and fast rule. There are variations in every product that influence its life and, while the values in Table 2.1 are typical lives, the distribution of failures almost always follows a curve such as that shown in Figure 2.2.

One of the difficulties in developing an effective failure analysis program is that it's easy to understand that "it shouldn't happen" when a shaft breaks but too often failures that don't involve catastrophic damage or an obvious fracture aren't recognized as failures. Two very common places where there is a lack of awareness of the need for failure analysis include corrosion and wear. These are applications where

TABLE 2.1

Expected Lives of Industrial Machinery

Component	Life[1] Based on Current Accepted Industrial Design Criteria
Anti-friction bearings	Generally over 20 years
Belt drives	
V-belts	2 to 3 years – MPTA standards say 25,000 hours
Synchronous belts	1.25 to 1.5 years – MPTA standards say 12,500 hours
Chain (power transmission)	2.5 years, but this is highly dependent on lubrication
Couplings	
Grid	Should never fail with good alignment and relubrication every 5 years
Gear	Should never fail with good alignment and relubrication every 3 years
Elastomer	5 to 10 years
Fans	
General industrial	5 to 8 years
Mine ventilation	15 to 20 years
Motors	
Less than 40 hp (30 kW)	7 to 9 years
40 hp (30 kW) and over	15 to 20 years
Plain bearings	
Smooth applications	Infinite – for equipment such as motors and generators.
Cyclical loading	Over 3 years in applications such as engines and crushers
Pumps	
End suction ANSI	5 years
Double suction horiz split	10 years
Reducers	Depends on service factor and actual loading
Below 1.5 SF	Typically 3 to 4 years
1.6 to 2.0 SF	4 to 8 years
Over 2.3 SF	Typically more than 10 years
Shafts[1]	Should never fail

Life is defined as "until major maintenance is required, i.e., overhaul, replacement, or rebuilding is needed".

[1] When we say "expected life", this is based on our experience with plants that we consider to have high-quality engineering and maintenance programs.

occasional replacement or refurbishment is needed but the rework frequency and costs are ignored until a crisis develops with the responsible personnel having lost site of the fact that their original assignment was to improve plant reliability and reduce costs.

(As every reliability professional has seen, another area where the value of failure analysis commonly isn't utilized involves the practice of routinely scheduled outages.

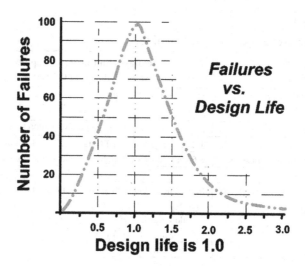

FIGURE 2.2   Typical industrial component life distribution.

Should maintenance be scheduled based on the position of the sun or the moon and end up with monthly or annual outages? What is the real difference between a scheduled outage and a breakdown? The need for maintenance has to be recognized, but in many cases the time between these "scheduled failures" can be greatly lengthened by analysis of the need for the various phases of the outage. We have never seen a situation where, after a competent and detailed analysis, the time between outages hasn't been increased by at least 50%.)

Diagnosing the Failure

The failed component *always* tells an accurate story about what happened. The problem with many analyses is that they succumb to human weaknesses and jump to conclusions before reading the entire story. A good failure analyst needs Sherlock Holmes' powers of observation along with an understanding of the operations surrounding the failure and the ability, diplomacy, and tenacity to question everything.

Looking at the failed component, or the fracture area, and its surroundings will tell almost everything about the physical causes of the failure, i.e., where it started, the magnitude and direction of the forces involved, how long it took to propagate across the piece, and whether there were any material weaknesses that contributed significantly to it. Sometimes it is more difficult than others to see the clues and sometimes it is difficult to understand all of them, but they are there.

Inspection of the failed agitator shaft shown in Photo 2.1 indicates that it was a fatigue failure, was primarily caused by two-way plain bending, had stress concentrations, and was lightly loaded. Further inspection of the shaft exterior surface will find evidence of the faulty maintenance practices that caused the two-way bending.

(Above we mention the subject of material defects. They are certainly important topics when dealing with failure analysis but are rarely significant contributors. There is a never-ending tendency to blame the materials, but our failure analyses have found a parent material problem with much less than 0.1% of the industrial

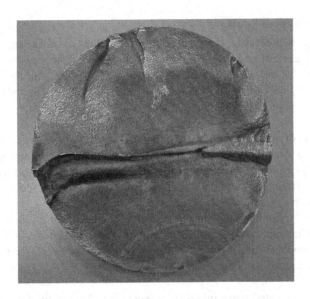

PHOTO 2.1   An agitator shaft fatigue failure.

failures. However, as soon as humans get their hands on a part, whether it is welding, heat treating, or fabrication processes, the proportion of failures increases.

(Most – certainly far more than 80% – of the failures found in the average plant can be solved without resorting to expensive metallography or hiring an outside expert. In these situations, an observant person, or preferably a team of persons, using good lighting and low power magnification, should be able to see enough of the features on the failed part to determine the type of failure and, from that, the forces that caused it. Then, with more investigation, the multiple human errors that culminated in the failure should be apparent.)

Over the past 40 years or so of looking at the major roots of failures, we have found the typical plant failure involves five +/– significant physical and human roots but more complex failures have more roots. The National Transportation Safety Board says aircraft crashes generally have 12 to 14 roots. Yet we frequently see people assigning one cause, one root, to a failure. Why? Our opinion it that is a combination of their lack of education and the fact that it is easy and a comforting thought that solving "that root" will get rid of the problem. (Looking at the latent roots certainly adds to the number of contributing causes.)

Failure analyses help the plant to find all of the things that contribute to a failure and then do something to prevent a recurrence. But we don't recommend that every failure be investigated and every root always be eliminated. That would be extraordinarily expensive and a waste of precious resources. What we would like to see people do is:

1. Recognize the significant failures and analyze their costs.
2. Identify all the roots, and then intelligently eliminate enough of the major ones to break that error-change linkage and ensure the problem never reappears.

Finding the Physical Roots

In conducting the analysis, it is critical that the analyst get to the site rapidly. As time goes on parts get lost and people's memories both fade or become colored, which makes finding the causes more difficult. As soon as possible, visit the site, inspect and collect the pieces, interview the involved personnel, and try to understand what happened. At the site:

- Take lots of photographs of the site and the pieces. Be sure there are enough photos to identify where the pieces started and where they ended up and try to always have some measurement scale in every photo. Also take photos of the surroundings and the associated parts, then take few more. They are inexpensive and frequently invaluable.
- Discuss the failure with involved personnel and ask them for their opinions about what happened. Try to get to them one or two at a time and ask them to explain what they think happened in their own words. (*Do not tell them something and ask them if they agree!*)

After you've completed these initial steps, step back, look at the failure and think about what could have caused it and what the possible ramifications might be. If there was or is the potential for a large economic loss, human injury, or a serious environmental event, then nothing should be spared in conducting a very detailed analysis. On the other hand, no organization has unlimited assets and an inexpensive failure deserves an inexpensive analysis.

If the failure is one of those with the possibility of major consequences, the analyst should follow a procedure such as the one below to find the causes:

1. Make a preliminary examination and take photos of the failed pieces. What type of failure was it? What basic forces were involved? What do the workers at the site think happened?
2. Discuss the failure with involved personnel who work with the equipment but were not there during the initial visit. Again, ask them for their opinions on the causes.
3. Collect background data. Look at the installation and the surroundings. What was the normal environment, power usage, operating conditions, etc? Also look at the same data for the machines that directly interact with the failed unit. Find out if there have been changes in the last month or year or ...
4. Under low power magnification look at the pieces, the fracture faces and other surfaces, develop a logic tree listing every symptom of the failure, and determine the possible causes of the symptoms and the failure scenarios.
5. Conduct more detailed analysis on the critical pieces. This may involve metallography, nondestructive testing, chemical analysis, residual stress analysis, or a variety of other inspection tools used to determine the state of the pieces.
6. From the above, determine the physical failure mechanisms and how the failure happened.
7. Then, go further and find the human and latent roots so you can determine how to prevent future occurrences.

Any of these steps can be eliminated or partially eliminated but each step that is skipped increases the chance of making an error. But again, the cost of the analysis has to be balanced against the true cost of the failure.

Of these seven steps, my opinion is that "Discuss with involved personnel" is definitely the most important point because we all have different viewpoints. The machine operators will see and hear things that the maintenance personnel don't recognize, the maintenance personnel will see things the engineers don't notice, the operating supervisor will see things the operator doesn't notice, and so on. Combining all of them will lead to a more comprehensive understanding of what went on.

(When you talk to these people, *listen* to both their words and to what they are trying to say. Many times, they don't have any technical knowledge and may not know the difference between a screwdriver and a hammer – but they have heard and felt things. We have heard pseudotechnical explanations that would make both Einstein and Newton turn over in their graves – but they provided great clues to the cause of the failures. Nevertheless, the eyewitness testimony has to be weighed against what the parts say and tempered by the knowledge that many people are well-meaning but not accurate in what they report.)

Comments on the Seven Steps – Continued

#1 – In most failure analyses identification of the failed part is relatively easy. But in complex failures, like an aircraft crash or the destruction of a large gearbox, finding the piece that failed first, usually by fatigue or wear, is sometimes difficult and every piece has to be looked at carefully.

#2 – In looking for changes don't be shortsighted. Sometimes they could have been made five or ten years ago. In one example we worked with a company that eliminated a dust collection system because they felt the annual maintenance was excessive. Ten years later, the resulting corrosion cost millions to repair. In a second application, the plant practice of leaving out $150 baffle gaskets in a group of heat exchangers resulted in corrosion that required that eight tube sheets to be replaced ten years later – at a cost of $80,000 each.

Photo 2.2 shows a pair of aluminum castings. The one to the left failed after the plant put a condensate hose on it to prevent it from freezing the product inside, (Why not repair the insulation?) and the condensate increased the casting temperature to over 80°C (180°F), well above the pitting corrosion threshold of the aluminum.

What are the sources of failures? Figure 2.3 is a tree that shows a breakdown of the four major failure mechanisms and their most significant contributors.

Looking at Figure 2.3:

- Corrosion is the probably biggest source of failures and costs most of the developed nations between 3 and 4% of their GNP, a number that has held fairly constant over the past 60 years. There isn't a lot of industrial data but a DuPont survey from 1968 to 1970 analyzed their corporate maintenance costs and found the following breakdown (but they said there was only one cause per failure):
  a) Corrosion – 55 to 60%
  b) Wear – 5 to 7%

PHOTO 2.2   Pitting corrosion from elevated temperatures on an aluminum housing.

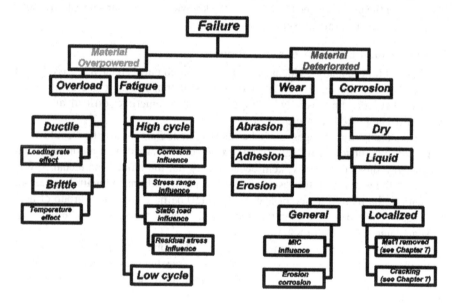

FIGURE 2.3   An expansion of Figure 2.1 showing the failure influences.

   c) Mechanical Failure
       – Fatigue – 90%+
       – Overload – remainder

- The chart doesn't specifically show a placement for the effects of thermal changes because temperature has a multitude of effects on materials and frequently has a huge effect on causing failures. Thermal stresses contribute to numerous failures yet rarely show any visible symptoms and we suspect that this lack contributes to the fact that many of the people designing and working with machinery don't understand the forces created by nonuniform heating. Two interesting examples involving temperature differentials and thermal effects are:

- Measurements on a shiny stainless steel tank roof found the roof temperature was about 55°C (100°F) cooler than the red oxide painted external roof beams. The stresses created during weather changes, e.g., a hot summer day with a sudden thunderstorm, were close to the yield strength of the metals and frequent cracking was the result.
- A plant repeatedly pumped 2°C (35°F) cooling water into a large vessel while the vessel wall was still at 95°C (200°F+). Eventually they had serious fatigue cracks.

But these examples just show the thermal stresses. In addition, it has to be recognized that variations in ambient temperatures can have a substantial effect on corrosion rates, can have a major effect on the impact strength of both metals and polymers, and will have an effect on chemical reaction rates. Moreover, large temperature variations can greatly affect both fatigue and wear mechanisms.

Introduction to Materials – Stresses and Strains

The next chapter will cover this in a lot more detail, but it is important to understand the type of physical force that caused the failure. An overload failure tells that something unusual has happened and usually results from a single event. In an overload failure the part either bends or breaks almost instantaneously. (We say "almost" because some ductile failures, such as the failure of an inexpensive wire coat hanger, involve the slow collapse of the piece.) On the other hand, fatigue failures result from multiple stress cycles and take place over a comparatively very long time.

The ability to understand the type of forces (the magnitude and the direction) gives tremendous clues as to the how those forces were involved. For example, if a conveyor link fractured the analyst would look at it and find:

- If it was an overload failure, such as the piece shown in Photo 2.3, the force was applied in the instant before the piece failed and some catastrophic event, such as the conveyor bucket snagging the side of the casing or some hard material, should be suspected. Also, the tiny thumbnail-shaped discoloration at the top would lead to some suspicion about the material condition.
- If it was a fatigue failure, the force that caused the failure was applied over many, many cycles and a process problem, such as a grossly improper flow rate, would be indicated as a major cause.

As mentioned earlier, from looking at the fracture face of the rotating agitator shaft shown in Photo 2.2 we can see that failure:

- Was a fatigue failure.
- Took a relatively long time.
- Was primarily caused by two-plane bending, but the loading also had a torsional fatigue component.

PHOTO 2.3   A chain link that has failed as a result of an overload. Note that the small dark area at the top of the photo was a crack that existed well before the catastrophic failure.

- Was very lightly loaded at the time of final fracture, but the load varied over time.
- Had an appreciable stress concentration on one side.

With this information we can first look at the rest of the assembly and ask how a rotating shaft can be stressed by two-plane bending during operation, then ask how those physical causes came to exist.

Sound interpretation of the surface features will yield a wealth of clues to the failure forces and the modifying conditions and will direct the analyst to the mechanism. From this point, finding the human errors that contributed to the failure can be accomplished with the advantage of a solid foundation.

Determining the Failure Mechanisms

One of the difficulties for a new failure analyst is deciding where and how to begin, i.e., how can they find the physical roots. With practice it becomes easier, but we almost always start by using the chart in Figure 2.4. (We realize that there is no substitute for experience and that this chart will not always lead to the answer. But we also know it works in better than 80% of the plant failures.) This is the first of several similar charts in the book as guides for the beginner – or a refresher for the experienced analyst.

The Plant Failure Analysis Laboratory

What tools are needed in a plant failure analysis laboratory? Below is a suggested list for a failure analysis lab. The items are listed in our opinion of the order of importance.

- *Area* – It doesn't have to be a large room and even a corner of an office will do in the beginning. What you need is a place to conveniently put the failed parts and examine them. Good, glare free, color corrected lighting is important. (There are three large shop lights with individual controls directly over our main examination bench.) Depending on the size of the failures in the plant, there may be a need for or access to material handling equipment.

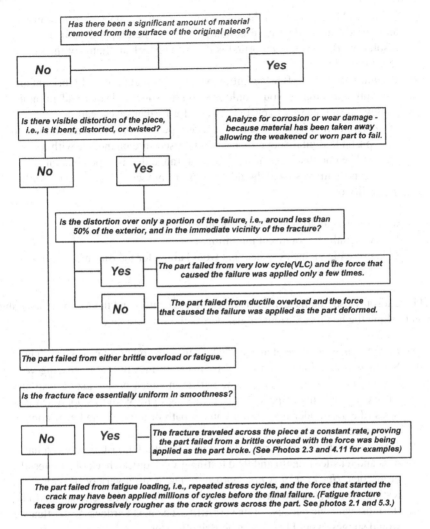

Has there been a significant amount of material removed from the surface of the original piece?

No

Yes

Is there visible distortion of the piece, i.e., is it bent, distorted, or twisted?

Analyze for corrosion or wear damage - because material has been taken away allowing the weakened or worn part to fail.

No   **Yes**

Is the distortion over only a portion of the failure, i.e., around less than 50% of the exterior, and in the immediate vicinity of the fracture?

Yes   The part failed from very low cycle(VLC) and the force that caused the failure was applied only a few times.

No   The part failed from ductile overload and the force that caused the failure was applied as the part deformed.

The part failed from either brittle overload or fatigue.

Is the fracture face essentially uniform in smoothness?

No   **Yes**   The fracture traveled across the piece at a constant rate, proving the part failed from a brittle overload with the force was being applied as the part broke. (See Photos 2.3 and 4.11 for examples)

The part failed from fatigue loading, i.e., repeated stress cycles, and the force that started the crack may have been applied millions of cycles before the final failure. (Fatigue fracture faces grow progressively rougher as the crack grows across the part. See photos 2.1 and 5.3.)

OVERLOAD failures happen as the load (force) is applied, i.e., the part breaks as it is being overstressed. A FATIGUE failure starts as a tiny crack and requires multiple stress applications before the part breaks.

FIGURE 2.4 Beginning the analysis, a logic tree that can be used to identify the type of failure and prepare for a more detailed diagnosis.

There should also be a storage facility where pieces can be kept for some time, at least a year and there should be treatment of steel part surfaces to prevent corrosion. (We have used barrister bookcases in several plants and positioned them in very visible locations. They serve as good advertising for the value of failure analysis and for the value of your services.)

- *Digital camera* – Two to four megapixels should suffice for most work, but even cell phones can be 12 meg and over. Of critical importance is that it has the ability to take close-up photos and that the area has good glare-free lighting. (Modern cameras have far more capacity than just four megapixels and an advantage they have is that photos can be cropped without losing resolution. But what is important is that you take clear photos of the important failure features.)

- *Cleaning ability* – After the initial examination and taking of lubricant and contaminant samples, you should have both a solvent cleaner and an industrial wash sink nearby to clean the parts for further investigation.
- *Microscopes and light* – The first microscope we would recommend would be a portable digital scope that could be used in conjunction with a computer. (The one that I use has a dual focal range that is a nice benefit.)

     The next microscope should be a basic binocular microscope with a capability of:

     a. About 7- to 20-power magnification.
     b. A digital camera direct mounting
     c. A reticule in one eyepiece that would allow for measurement of magnified sections.

The light source should have a variable brightness control and flexible head to enable you to shine the beam at various angles across the fracture face.

- *Lazy Susan* – For ease of analyzing bearings and other rotating parts, a lazy Susan is invaluable. It can easily be made from a turntable bearing with a round plate mounted on it. (The part is mounted on the lazy Susan and rotated while being inspected with a varying light. With this approach features like a meandering bearing contact path or a misaligned gear contact pattern can be easily seen and measured.)
- *Hardness testing equipment* – There is a direct relationship between hardness and tensile strength and field testing gives a quick check of a material's properties. *However, it is not foolproof!* There are a variety of scales and testing devices, but we have found an Equotip electronic scleroscope to be the most valuable because of its versatility. The next hardness tester that we would suggest would be a portable Brinell tester.

(In discussing the equipment needs for a plant FA laboratory we've assumed the plant has an active Predictive Maintenance Department with tools such as a vibration analyzer, strobe light, infrared scanner, etc.)

When you start the program, we suggest having the metallurgical work conducted by an experienced outside contractor. There are several reasons for this approach:

1. Metallurgy is extraordinarily complex and, unless you have years of education and experience in that field, it is best left to a professional.
2. Metallurgical analysis will be needed on all failures where there is a chance of serious injury or a significant environmental event and you will want the analysis to be able to withstand professional scrutiny.
3. Working with metallurgists and developing a rapport with them will help you to better understand how the material's properties and structure affects failures.

CHAPTER 2 SUMMARY

There are only four basic failure mechanisms: overload, fatigue, corrosion, and erosion. All failures fall into these categories.

Deciding how deeply to go into a failure analysis has to be made early in the analysis and should be based on safety and cost.

The part will explain how and why it failed. The difficulty is in interpreting the failure clues.

Overload failures happen instantly as the force is applied. Fatigue cracks require repeated force applications and may take millions and millions of load cycles to progress across the part.

A maintenance group is really good at replacing or rebuilding is needed. something they've lost sight of their primary goal and it's time to do a failure analysis.

BIBLIOGRAPHY

Case Histories in Failure Analysis, American Society for Metals, Metals Park, OH, 1979, ISBN: 0-87170-078-6.

Latino, Robert J., Latino, Kenneth C., and Latino, Mark A., *Root Cause Analysis*, CRC, 2011, ISBN: 978-1-4398-5092-3.

Metals Handbook, Volume 1, Properties and Selection, American Society for Metals, Metals Park, OH, 1990, ISBN: 978-0-87170-377-4.

Metals Handbook, Volume 8, Failure Analysis, American Society for Metals, Metals Park, OH, 1979, ISBN: 978-0-87170-389-7.

Wulpi, Donald, *Understanding How Components Fail*, ASM, Metals Park, OH, 1986, ISBN: 0-87170-189-8.

3 Materials and the Sources of Stresses

A. Stress, Strain, and Material Strength
B. Mechanical Stress
C. Thermal Stress

Some very basic failure analyses can be accomplished without truly understanding the effects of various stresses and the materials they act on. Unfortunately, as failures become more complicated, this lack of knowledge almost always leads to serious errors and, when one starts looking into more costly or more complex failures, the need to understand exactly what is happening is incredibly valuable.

Stress causes parts to bend or break, but all materials don't react in the same manner. In the real world, a good understanding of stresses and their effects on materials is crucial to accurate failure analysis. (For example, doubling the load on a beam could cause it to deflect twice as much or it could cause an immediate catastrophic failure. But the effect of doubling the load on a rolling element bearing is that its life will be cut by a factor of eight to ten.) In this chapter, we'll discuss some of those basic definitions and concepts.

STRESS

When a load is put on a part, the part is *stressed*. Figure 3.1 shows a block with a cross-section area of 645 mm^2 (1 in^2). With the load of 50,000 N (11,240 pounds), the resultant tensile stress is 77.5 MPa or 11,240 lbs/in^2.

In this example, the block is axially loaded and the piece is in *tension*. In a similar manner, reversing the forces shown here would put the piece in compression. To calculate the magnitude of the stress in the block shown in Figure 3.1 the formula

$$s = p/a$$

is used. In this equation s is the pressure or stress (in MPa or pounds per square inch), p equals the load (in newtons or pounds-force), and a is the stressed area (in square mm or square inches).

If the same block were to be subjected to a bending load or a bending *moment* the stress would be very different as seen in Figure 3.2. The block in Figure 3.1 is subjected to pure tension and the stress is uniform across it but, as shown in Figure 3.2, bending

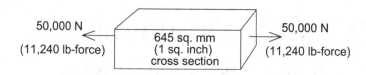

FIGURE 3.1   A basic example of tensile stress on a block.

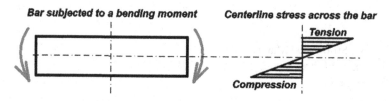

FIGURE 3.2   An example of a bending stress, usually called a *bending moment*. The figure to the right shows the stress profile at the center of the bar, and we can see it is zero along the horizontal centerline.

puts compression on one side of the piece and tension on the other. Calculating the stress resulting from the bending, the formula

$$s = Mc/I$$

is used where M is the bending moment, c is the distance from the centerline to the outer edge, and I is the moment of inertia of the piece.

In addition to tension and bending moments, the piece could also be loaded in a twisting manner with a *torsional* stress, or non-planar forces that would put a *shear* stress on it, or even a combination of stresses.

During the design of a part, these loads and the resultant stresses are usually calculated very carefully, and then a significant safety or service factor is added to guard against errors and unexpected failures.

One of the valuable features of looking at a failed part is that fractures are always perpendicular to the plane of the maximum stress. At the very beginning of a failure, when only a few grains are involved and the crack is too small to be visible to the unaided eye, the crack may not be perpendicular to the plane of the stress. But once the crack gets big enough to be seen, it always grows perpendicular to the stress and this is invaluable for its ability to point to the source of the load that caused the failure. Examples of this analysis of the fracture plane describing the major load direction are shown in Figure 3.3.

ELASTICITY

To some degree, all materials are *elastic*. This means that, as a load is put on the part the part stretches, as though it were a rubber band. The amount of stretch is directly proportional to the load up until the *yield point* (or *yield strength*) of the material and the engineering term for this elongation (stretch) is *strain*. Therefore, if a part is stressed, and the load remains in the elastic range for the material, the part will spring back into its original shape when the load is released.

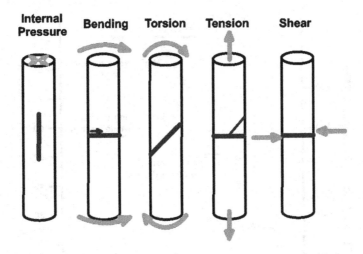

FIGURE 3.3   Except for pure shear, the fracture plane is always perpendicular to the force (the stress) that caused it.

PLASTICITY

But if the part is stressed beyond the yield strength, permanent deformation takes place and the piece becomes *plastically* deformed. (An example of this is an over-stretched paper clip that is plastically deformed and can no longer retain the papers.) If still more load is placed on the part, it will continue to plastically stretch and eventually fracture when the *tensile strength* is exceeded.

Figure 3.4 is a stress-strain diagram for two steels. The lower solid line shows a mild steel with a yield strength of 310 MPa (45,000) psi and a tensile strength of about 410 MPa (60,000 psi). This means that, if the part has a cross-section area of 645 mm^2 (1 in^2), a load of 20,411 kg (44,999 lbs) can be put on the part before it permanently deforms by 0.2%. If the load on the part is further increased and the yield strength exceeded, plastic deformation takes place. Additional loading causes more deformation and the part stretches until it breaks. Next, look at the dashed high strength steel line and notice that, up to the yield points, the metals elastically deflect at the same rate. The difference between the two alloys is that the high strength steel can carry a much higher load before it permanently deforms, i.e., before the yield strength is exceeded.

Confusing many people is the fact that the lower strength material is almost always more ductile and usually has a much greater distance between the yield and fracture points.

(The engineering *yield strength* of a material describes the point where the material has elongated by 0.2% under a given load. This can cause a misunderstanding in some applications because the material will actually plastically deform at stresses less than that, but that deformation is less than the critical 0.2%. In most applications, the difference isn't significant. But if there is a part such as a fastener or a piece of a machine structure that is stressed to the point where it yields by 0.01% on each stress cycle, and that occurs many times, it can contribute to a failure. Very low cycle [VLC] fatigue loads are often in this category.)

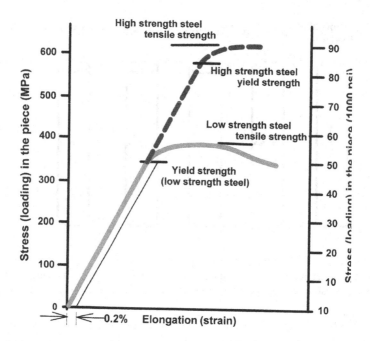

FIGURE 3.4 A typical stress-stain diagram showing how much a part distorts with a given load (stress).

From an elongation and plasticity viewpoint, the major difference between ductile materials and brittle materials is that the brittle materials have an extremely short range between their yield and tensile strengths and in some materials, like glass, it is essentially zero. On the other hand, some very ductile materials can be stretched until they double in length without fracture.

Most engineered designs use ductile materials and use the yield strength as the primary design criteria, applying an appropriate safety factor. The designer knows that a deformed part, a part where the load has exceeded the yield strength, is no longer serviceable and rarely is concerned about the load when the piece breaks in two. (Tensile strength is commonly used as a design criterion with brittle materials.)

Photo 3.1 shows a miniature tensile test specimen that was cut from a failed piece of equipment and was used to determine the actual properties of the steel. Tensile testing can determine the strengths and the elongation of the material. In the tensile

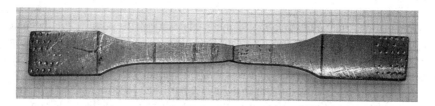

PHOTO 3.1 A deformed miniature tensile testing specimen showing the gage marks and the deformation typical of a ductile metal.

testing machine, the jaws are locked onto the test piece and the sample is slowly and steadily stressed in tension. The yield strength is determined by the applied stress when the separation between the two jaws has increased by 0.2%. The tensile strength is determined when the piece actually fractures and measuring the change in distance between two gage points on the bar after the fracture shows the elongation.

MODULUS OF ELASTICITY (YOUNG'S MODULUS)

In Figure 3.4 one can see that even though they are different steel alloys, up to the yield strength they deflect the same amount with equal loads. This property, the amount of deflection for a given load, is the *modulus of elasticity,* and is frequently called *Young's Modulus.* For all carbon steels, whether the weak inexpensive stuff paper clips are made from or a super strong heat-treated alloy, the modulus of elasticity is about 30×10^6 and they elongate identically. (The reason we sometimes use higher strength materials is that they will elastically deform much more before taking a permanent set, i.e., before their yield strength is exceeded.)

Other metals have differing moduli of elasticity as shown in Table 3.1.

Looking at the modulus of elasticity values in the chart, one sees that if identical pieces are made out of aluminum and steel and loaded in the same manner, the aluminum part will elongate or deflect almost three times as much as the steel one. By the same token, a copper piece with the same load would deflect about twice what the steel piece would do. This elongation (or *strain*) is relatively easy to calculate.

The modulus of elasticity, E, is expressed in GPa or psi and can be visualized as the load that would cause a perfectly elastic part to double in length. Knowing the stress (s), area (a), part length (l), and Young's modulus (E), it is relatively simple to calculate the elongation or strain with the formula:

$$\text{elongation} = \text{stress} \times \text{length} / \text{Young's modulus}$$

TABLE 3.1

Elastic Moduli (Young's Modulus) for Some Common Materials

Material	Metric Units – GPa	US Units – $\times 10^6$ psi
Aluminum	69	10
Beryllium	269	39
Cast Iron	90 to 150	13 to 22
Copper	105	15.5
Lead	14	2
Monel	175	25
Stainless Steel	190	28
Mild Steel	206	30
Titanium	103	15
Nylon	7.5 to 19	1.1 to 2.8
Polyethylene	48 to 95	7 to 14

So, if we go back to that example in Figure 3.1 and the steel block is 230 mm (5.84″) long we can calculate the elongation (ε).

$$\text{Elongation} = 11{,}240 \text{ psi} \times 5.84'' \, / \, 30{,}000{,}000 = 0.0022''$$

TOUGHNESS

Above we discussed the concepts of tensile and yield strengths. In conducting the laboratory tests to find these strengths, the test samples are clamped in a large machine and the load on the sample piece is very slowly increased. Running the testing machine may take over a minute starting with no load until the time a part is torn in two. But, in the real world there are times when loads are applied very rapidly, i.e., the parts are impact loaded such as when something is struck with a hammer. It is important to realize that there is very little connection between *strength*, with slowly applied loads, and *toughness*, with very rapid load application.

For instance, look at the two bars of the same size shown in Figure 3.5. The bearing steel bar is almost 10 times as strong as the mild steel bar. But what would happen if two rings, 75 mm × 12 mm × 3 mm (3″ diameter × 1/2″ wide × 1/8″ thick), were made, one from the mild steel bar and the other from the heat treated bearing steel, and both were struck hard with a large hammer? The mild steel ring would be badly

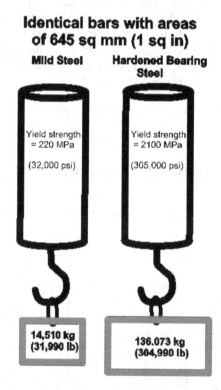

Identical bars with areas of 645 sq mm (1 sq in)

Mild Steel	Hardened Bearing Steel
Yield strength = 220 MPa (32,000 psi)	Yield strength = 2100 MPa (305,000 psi)
14,510 kg (31,990 lb)	136.073 kg (304,990 lb)

FIGURE 3.5   Yield strength examples.

bent but the hardened steel ring, even though it is almost 10 times stronger, would shatter like glass. The mild steel part has the ability to absorb more energy and is much tougher than the stronger bearing steel.

This *toughness* is the ability of a part to absorb energy while fracturing. There are many different toughness testing methods and one of the common ones is an Izod test, and Figure 3.6 shows a sketch of an Izod tester in operation. (A Charpy test is similar except that the vise that holds the part is positioned differently.)

The sample is mounted in the vise. Then the hammer is raised to the starting height and allowed to swing freely, hit and break the sample, and continue up to the finish height. For most samples the hammer would break the bar and swing up to a point a good deal lower than the starting height. This difference in height is carefully measured and multiplied by the mass of the hammer to determine exactly how much energy the sample piece absorbed. If the part were brittle, like glass, it would absorb almost no energy and the hammer's starting and finish points would be almost identical. At the other extreme, some tough nickel alloys are known as "hammer stoppers" and they need a tremendous amount of energy to fracture.

In the temperature range of −75°C to +350°C (−100°F to +650°F) tensile and yield strengths for most steels are relatively constant, changing by no more than 10%. But for many materials, both metals and plastics, the toughness varies tremendously, sometimes by a factor of more than ten, between about −25°C and 100°C (−10°F and 212°F) and some examples of this change in toughness with temperature are shown in Figure 3.7.

(The reason for this change in toughness is that the fracture mechanism and the energy needed to propagate the crack can change with temperature. For example, with a 0.45% carbon steel, at temperatures over 70°C [~160°F] the fracture mechanism is usually *ductile rupture*, where it takes a great deal of energy to tear each metal grain apart. As the temperature decreases the fracture mechanism gradually changes to cleavage along the grain boundaries and requires much less energy.)

So, when we look at the designed loading on a part, it could be a relatively steady load, like a building column, or it could be an impact load, such as a hammer blow,

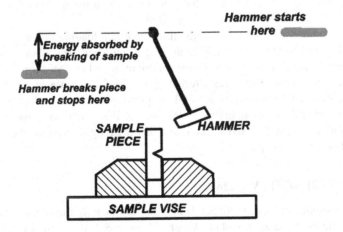

FIGURE 3.6   An Izod impact tester.

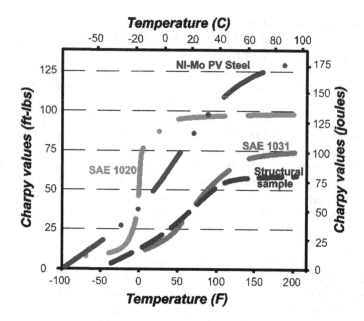

FIGURE 3.7 This shows how the impact strength of pressure vessel, structural (A 36), and some low carbon steels vary with temperature. Alloying elements can greatly affect the amount of change and the temperature sensitivity.

or it could be cyclically (fatigue) loaded, like a connecting rod in an engine, and the three types of loading require very different approaches.

FATIGUE

Long ago, in the early days of metal components, people recognized that there were some failures that didn't happen instantly from overload or impact loading. These failures occurred over time in parts that were in constant use and, relating them to tired human muscles, the term *fatigue* was used.

Figure 3.8 shows a comparison of the three common loading methods. As mentioned earlier, yield and tensile strengths are important when the load on a part is relatively steady. Impact strength (toughness) is important when there is a shock load, and one sees that fatigue involves alternating tensile forces on a piece. Typically, the long-term *fatigue strength* of a metal is approximately one half of the tensile strength, however there are many variables, such as component size, force application frequency, force range, and so on, that affect the fatigue strength and these are reviewed later in this chapter.

FATIGUE STRENGTH VS. TIME

The fatigue strength of a metal depends both on how heavily the part is loaded and how many times the load is applied. Figure 3.9 is an S-N diagram, where S is the stress and N is the number of load cycles, and it shows how fatigue strength changes

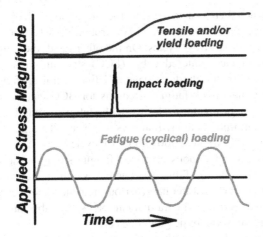

FIGURE 3.8 This shows the three different loading (force application) situations and the text explains how different strengths are required

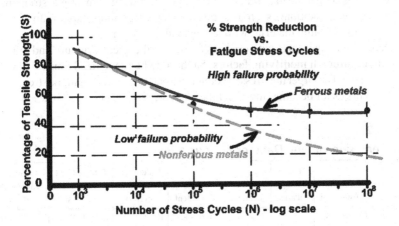

FIGURE 3.9 An S-N diagram showing how the fatigue strength of a metal varies with the number of stress cycles.

with the number of cycles. In the S-N diagram, it can be seen that, if a part is heavily loaded it can only withstand a relatively few fatigue stress cycles and as the load decreases, the number of stress cycles before failure increases.

This is an essential diagram and there are several points about it that should be discussed in much greater detail.

1. S-N charts usually show fatigue strength as though it was a very well-known and precise property when in fact the range between 1% survival and 99% survival is frequently an order of magnitude, i.e., one part may fail in 10,000 cycles while another apparently identical part with an identical load might last 100,000 cycles.

2. Generally, when there is no corrosion, the fatigue strength of steel and other metals that have a body-centered cubic structure (BCC) does not decrease after about 10^7 or so stress cycles. On the other hand, the fatigue strengths of metals with face-centered cubic (FCC) structures decrease continuously and FCC materials don't have a defined limiting fatigue strength. Consequently, handbook fatigue strengths for BCC metals, such as steel, are given for 10^7 or greater cycles while the fatigue strengths for FCC metals (copper, aluminum, etc.) will always be given with a qualifying number of cycles, e.g., 100 MPa (14,500 psi) @ 10^6 cycles.

3. Corrosion has a tremendous effect on fatigue strength and continuously reduces it with time. In Table 3.2, one can see that even material seemingly as benign as river water has a serious detrimental effect on the fatigue strength. There is no good data that can be used to absolutely predict this deterioration, but some basic guidelines are:

   a. If there is visible corrosion or rusting the fatigue strength has been reduced.

   b. The longer the corrosion goes on, the more the fatigue strength is reduced.

   c. The stronger the material, the more rapid the reduction in fatigue strength.

   d. Aerated solutions, such as sprayed water, cause more rapid reductions than the same material with less available oxygen.

4. Fatigue damage starts with changes to the metallurgical structure and other fatigue strength modifying factors that have to be recognized are:

   a. Size – The larger the stressed volume of a part the lower the fatigue strength. This factor becomes more significant as the component lives grow longer.

TABLE 3.2
Some Corrosion Fatigue Data for Ferrous Alloys

Material	Heat Treatment	Tensile Strength MPa/ksi	Type of Loading	Corrodent	Endurance Basis (cycles)	Fatigue Limit in air MPa/ksi	Corrosion Fatigue Str MPa/ksi
Mild steel	Normalized		Rotating Bending	River water drip	10^8 cycles	234 MPa/34ksi	28 MPa/4.1ksi
0.21% C steel	Annealed	438/63.6	Torsion	Sea water	10^8 cycles at 1300 cpm	197/28.6	26/3.8
0.21% C steel	Annealed	438/63.6	Torsion	Sea water	10^8 cycles at 1500 cpm	124/18.0	34.5/5.0
12.5% Cr steel		895/130	Torsion	Fresh water	2.5×10^6	225/32.6	110/16.0
18/8 SS		1150/168	Torsion	Fresh water	2.5×10^6	170/24.6	73/10.6
18.5% Cr	Annealed	690/100	Torsion			215/31.2	170/24.6
0.5 C, Cd plate		990/144	Torsion			178/25.8	45.5/6.6
SAE 1035	Normalized	540/78	Rot bend	6.8% salt sol'n	10^7	249/36.2	152/22.0
SAE 1035	Normalized	540/78	Rot bend	6.8% salt sol'n +H_2S	10^7	249/36.2	65/9.4

From Osgood, C. C., *Fatigue Design*, Wiley Interscience, 1970.

b. Loading Type – Axial and torsional fatigue strengths are less than bending fatigue strengths, primarily because they involve a larger volume of the part being stressed.

c. Surface Finish – Surface roughness becomes more important as the strength of the parent material increases, but under any circumstances poorly machined or cut surfaces will reduce the fatigue strength of a part. The rougher or more irregular the surface, the lower the fatigue strength

d. Surface Treatment – Processes that leave the surface with a compressive stress, such as case hardening, shot peening, and planishing, will increase the fatigue strength of the part. On the other hand, surface finishes like hard plating can substantially reduce it.

e. Temperature – Above about 320°C (~600°F) the fatigue strength of steels starts to decrease significantly.

Therefore, when calculating the actual fatigue strength of a material, i.e., the fatigue strength of the part in operation, one has to use:

Actual Fatigue Strength = Textbook fatigue strength × Size factor × Loading factor × Surface finish factor × Surface treatment factor × Temperature factor × Corrosion factor

In addition, it is well recognized that low frequency fatigue stress applications will reduce the fatigue strength more rapidly than high frequency stress applications.

SOME BASIC METALLURGY

Most of the metal used in the structure and equipment today's manufacturing plants is relatively low strength steel. This *mild steel* is an alloy of iron and carbon. If it were just iron it would be extremely weak, so carbon, up to about 0.3%, is added to strengthen the alloy but there is little else in the way of alloying elements.

Figure 3.10 shows a two-dimension view of the basic structure of an iron crystal. (In reality, the same structure would be repeated almost endlessly in the third dimension and there would be many other crystals, most in other orientations.) Each of the "Fe" symbols represents an iron atom and, because it is a metal, the crystal structure is regular and uniform, but the structure is very weak and the rows of atoms can relatively easily be slid over each other, a little like a deck of cards.

In Figure 3.11 some carbon atoms have been added to the iron and their addition makes steel. Carbon is an *interstitial atom* and they strengthen the steel by reducing the ability of the iron atoms to slide over each other.

Later we'll discuss some other alloying elements and how they can further improve the steel's properties. But, unfortunately, there are times when too many additional atoms are added to the structure and the result is a general weakening. Looking at Figure 3.12, it is relatively easy to see that opposing vertical stresses across the plane of the impurity would readily lead to a failure.

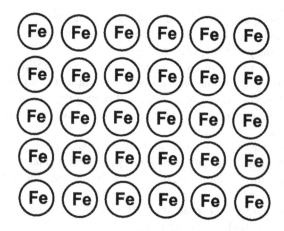

FIGURE 3.10   A two-dimensional sketch of the layout of atoms in steel.

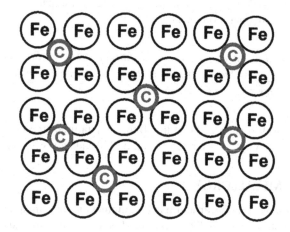

FIGURE 3.11   Adding carbon to iron greatly strengthens it.

There are some impurities that are intentionally added, and some are the result of processing errors and product contamination. For example, there are times when sulphur or calcium is added to allow more rapid and less expensive machining of steel pieces. One of the negative effects of these additions is that, although they don't measurable affect the tensile and yield strengths, they invariably reduce the fatigue strength of the pieces, typically by about 15%.

With the knowledge that the basic structure of steel and almost all other metals consists of a metal crystal structure strengthened by alloying elements, one can then begin to look at the major variations and their results.

CARBON STEELS

Plain carbon steel is an alloy of carbon and iron. The largest portion of this mix is iron, but as shown above, without the addition of small amounts of carbon, iron is extremely weak. Carbon acts as a strengthening agent and the amount of carbon can

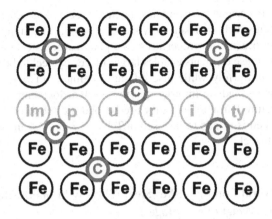

FIGURE 3.12 Adding an impurity to the steel structure can weaken it.

vary from about 0.06% up to about 1.2%. The real value of the carbon is that it gives steel a huge improvement in mechanical properties compared to those of the basic iron. Very low carbon steels such as SAE 1010 are relatively weak, while higher carbon steels such as SAE 1045 or 1095 are harder and stronger, and above about 0.3% carbon, they can be heat-treated to even greater strengths.

A general division of these carbon steels is:

- *Low carbon* (often referred to as *mild steel*) – up to 0.29% carbon
- *Medium carbon* – from 0.3% to 0.6% carbon
- *High carbon* – from 0.6% to 1.1% carbon

(Between 1.2% and 1.7% C is a material, usually called semi-steel, that shares many of the properties of cast iron and is commonly used in the cast form.)

In addition to carbon, some of the usual elements in low carbon steel are silicon, manganese, phosphorus, and sulphur. The silicon, up to 0.6%, and manganese, up to 1.65%, are added to help remove oxygen from the molten mix. (If the oxygen isn't removed it combines with other materials and forms oxides, particles that weaken the finished product.) The phosphorus and sulphur are contaminants and their levels are closely controlled.

The advantages of plain carbon steel are that it:

- Is relatively low cost.
- Does not take very sophisticated equipment to make.
- Can readily be worked (or formed).
- Offers a great improvement in strength and ductility over wrought iron, cast iron, and most of the nonferrous alloys.

The problems are that it:

1. Is not as strong as we would like it to be.
2. Weakens rapidly at higher temperatures.
3. Cannot be heat-treated to be hard and strong without becoming very brittle.
4. Doesn't have very good corrosion resistance.

Low alloy steels are designed to overcome the first three of those objections and have additions of elements such as nickel, molybdenum, boron, vanadium, and chrome, up to a total of about 6%, so the properties of the plain carbon steel can be improved. Generally, adding these elements enables larger pieces of steel to be hardened more successfully and uniformly while also adding toughness. (There are also special additions that can be made to improve qualities such as low temperature toughness and high temperature hardness, but they are beyond the scope of this text.) Below we'll define the chemistry and terms used in Tables 3.3 and 3.4 and discuss some of the basic types of steel.

TABLE 3.3
Showing Several Common SAE* Steel Alloys with Major Chemical Constituents (Weight Percentages)

SAE#	Carbon	Manganese	Chromium	Nickel	Molybdenum
1018	0.15–0.20	0.60–0.90	nr	nr	nr
1020	0.18–0.23	0.30–0.60	nr	nr	nr
1035	0.31–0.38	0.60–0.90	nr	nr	nr
1045	0.42–0.50	0.60–0.90	nr	nr	nr
4140	0.38–0.43	0.75–1.00	0.80–1.10	nr	0.15–0.25
4340	0.38–0.43	0.60–0.80	0.70–0.90	1.65–2.00	0.20–0.30
8620	0.18–0.23	0.70–0.90	0.45–0.60	0.40–0.70	0.15–0.25
52100	0.98–1.10	0.25–0.45	1.30–1.60	nr	nr

SAE = Society of Automotive Engineers, the technical society that develops the chemical standards for the common steel alloys. For many years the American Iron and Steel Institute (AISI) used the SAE designations, and it is common to hear alloys referred to as "AISI 1018" or "AISI 4140" but the responsible agency is the SAE.

TABLE 3.4
Reviewing the Typical Physical Properties of Some of Alloys with Different Processing Methods

SAE#	Condition (see below)	Tensile Strength MPa/ksi	Yield Strength MPa/ksi	% Elongation	Hardness BHN
1018	Hot rolled	380/55	220/32	25%	115
1020	Hot rolled	380/55	205/30	25%	115
1035	Hot rolled	495/72	275/40	18%	145
1045	Hot rolled	565/82	310/45	16	160
1045	Cold drawn	625/91	530/77	12%	180
1045	Q & T	825/120	620/90	18	240
4340	Q & T	1250/180	1100/160	15%	360
8620	Hot rolled	615/89	450/65	25%	188
8620	Cold drawn	705/102	585/85	22%	210

ASM Handbook, 8th edition, Volume 1, 1961, and the Ryerson Stock List, Joseph T. Ryerson & Son (a subsidiary of Inland Steel), 1995.

In a very simple comparison, making a metal alloy like steel is a little like making bread; you assemble the ingredients, mix them in a certain way, then cook them for a given time. There are the essentials, like flour and liquids (iron and carbon), then a number of additives that give the mix a special flavor, like raisins, maple syrup, chocolate, and so on (nickel, chrome, molybdenum, etc.). The way that the bread is kneaded changes the end result, as does the way that it is heated. With steel, by mixing the correct additives at the proper time and using either heat or cold working techniques to strengthen the alloy, a tremendous range of strength and ductility properties can be controlled to meet our needs.

In analyzing the choices as to what alloy should be used, the component strength isn't the only thing to consider. In addition, size, final cost, ease of fabrication, reliability, and so on, all have to be looked at. If all this sounds somewhat confusing to you, you're not alone and shouldn't feel badly. In one of the books in our library there is a listing of sun gears from a series of automotive automatic transmissions. One would think there wouldn't be a big variety in the materials chosen because of the similarity of the applications. However, there were 7 manufacturers, 11 transmissions, 9 different steels, 2 heat treatments, and 10 different hardness specifications. Three of the manufacturers had more than one transmission and in *no case* were the specifications the same. Metallurgy and materials science are very complicated and if you feel a little confused it seems like you have good company.

IRON AND ITS ALLOYING ELEMENTS

Before we go further, two of the truly important concepts to understand are *hardness* and *hardenability*. There is a direct relationship between hardness and tensile strength and the harder a part is, the stronger it is. This property of hardenability describes how readily a steel can be hardened and, after cost, really drives much of material selection. It's important to realize that hardenability involves both the depth and the distribution of the hardness that results when a part is heat treated and many of the alloying elements mentioned below are used to improve hardenability. (But in all this discussion, don't forget the difference between strength and toughness. Very hard materials are frequently brittle and most have poor toughness. An example of this can be seen by looking at high-strength ceramics that are hard but typically have poor impact properties.)

The principal alloying elements for iron and some common applications are:

- *Carbon* converts pure iron to steel. Up to a little over 1%, the higher the carbon percentage, the harder and stronger the steel. Above about 1.7% carbon the material is usually called *cast iron.*
- *Chromium* increases the ability to heat treat steel and increases the depth of hardness. From:
  - 0.3% to 1.6%, it is used to improve heat treating response in low alloy steel
  - 10% to 27% it forms stainless steel. (There are some inexpensive grades of stainless steel with less than 10% chrome and marginal corrosion resistance.)

- *Manganese* deoxidizes (removes oxygen from) the mix and also improves response to heat treatment.
  - 0.3 to 1.9% – Low alloy steels
  - 10 to 15% – Work hardening steels (Hadfield steel)
- *Molybdenum* increases toughness and depth of heat treatment
  - 0.08 to 0.6% in carbon and low alloy steels
  - 0.6 to 4.0% in stainless steels, where it also used to increase pitting resistance with up over 6% in some specialty stainlesses.
- *Nickel* increase strength, toughness, and impact resistance
  - 0.4 to 3.75% in low alloy steels
  - 3.5 to 36% in stainless steels and corrosion resistant alloys (CRAs)
- *Silicon* deoxidizes and improves tensile strength.
  - Normally in the range of 0.2 to 0.35% in low alloy and carbon steels but it can range up to 1%.
  - From 1.8 to 2.2% in specialty silicon manganese and low permeability electrical steels.

Important concept – As shown in Figure 3.13, many of these elements improve the response to heat treatment and this means that the hardened portion of the steel extends to a greater depth. For instance, if a 75 mm (2.95″) diameter plain 0.4% carbon steel bar is heated in a furnace, and then water quenched, the result is a hardened layer only 3 mm (1/8″) deep. But if the bar is a low alloy steel, such as SAE 4340, with chrome and nickel in it, the hardened area will extend all the way to the center of the bar and the resultant structure will be both stronger and tougher.

(Other alloying elements sometimes seen include aluminum, boron, calcium, columbium, copper, lead, sulphur, tellurium, titanium, tungsten, and titanium. Lead, calcium,

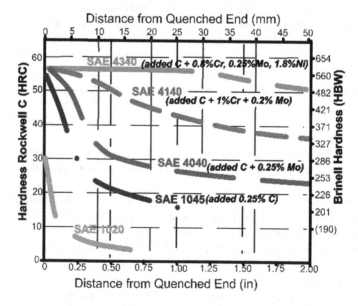

FIGURE 3.13 Showing how the addition of various alloying elements allows the hardness to penetrate deeper into a test bar. This carries over to all quenched steel parts.

and sulphur don't really alloy with the iron, but form long "stringers" that help the steel to machine more easily. The others are used to improve specific qualities in the steel.)

SAE NUMBERING CODE

In North America, the Society of Automotive Engineers (SAE) system is commonly used to identify carbon and low alloy steels. There is another system called the Unified Numbering System (UNS) code which can also be used. The UNS system is built around the SAE system for common steels and it also is used for the identification of many other metals. There will sometimes be references to American Iron and Steel Institute (AISI) steel numbers, for instance, AISI 1045, but the AISI uses the SAE system as its base and the correct designation would actually be SAE 1045.

The SAE system uses a four- or five-digit number to identify a steel and understanding the sequence will help describe the steel constituents. For example:

- SAE 1045 – The first two digits ("10") show it is a plain carbon steel, while the last two ("45") state that it contains 0.45% carbon. (Including tolerances, the allowable carbon range is actually 0.42 to 0.50 weight percent.)
- SAE 4130 – The first two digits ("41") indicate there are chrome and molybdenum additions. The "30" shows that the steel has been alloyed with 0.30% carbon.

The most common ranges of these SAE steels are:

10xx – Plain carbon steel, non-resulphurized
11xx – Plain carbon steel, resulphurized
12xx – Plain carbon steel, resulphurized and rephosphorized
41xx – Low alloy steel, chromium (0.50, .80, or .95%) and molybdenum (0.12, 0.20, or 0.30%) additions
43xx – Low alloy steel, nickel (1.83%), chromium (0.50%), and molybdenum (0.25%) additions
46xx – Low alloy steel, nickel (0.85, 1.83%), and molybdenum (0.20, .25) added
5xxxx – Specialty bearing steels, carbon (1.04%), chromium (1.03 or 1.45%)
61xx – Low alloy vanadium steel, chromium (0.60 or .95%), vanadium (0.13% minimum)
86xx – Low alloy steel, nickel (0.55%), chromium (0.50%), and molybdenum (0.20%) additions
xxBxx – Indicates a specialty steel where boron is added to improve the hardenability of the steel and the action of the other alloying elements

(There are many other numbering systems. For example, most of the major industrial countries such as Great Britain, Japan, and Germany have extensive systems for identifying materials. In the United States, the ASTM has a series of standards that cover everything from boiler tube materials to tennis shoes. The ASTM standards are generally application specific such as steel for pressure vessels and bolting for high temperature applications. Many materials are formulated to meet several different organization's standards so a plain carbon steel might meet an SAE specification and also meet ASTM specs as well as DIN, ISO, and JIS standards.)

Plain Carbon Steel – The basics

1010, 1018, 1020 – Low carbon (mild) steels – Low strength, used for general
   structural and mechanical parts where a "steel" is needed, also frequently
   used for case hardening.

1035, 1042, 1044, 1045 – Medium carbon, direct hardening steels used where
   stronger, more wear resistant materials are needed.

1095 – High carbon direct hardening steel, frequently used for drills and hardened tools.

Free Machining Steels – These steels have tensile and yield properties similar to their
"plain carbon steel" counterparts but are made with additives that make machining
easier and less expensive. From a reliability viewpoint, one problem with these steels
is that they usually don't have as good fatigue properties as the plain carbon variet-
ies, because the same features that make them free machining and allow the chips
to break up easily also cause small stress concentrations. Their fatigue strength is
frequently decreased by 10 to 20%.

1117 – Free machining, resulphurized steel, frequently used for case hardening

1137, 1141, 1144 – Medium carbon, direct hardening, resulphurized free
   machining steels

1214 – Low carbon screw machine stock, resulphurized and rephosphorized

Low Alloy Steels – Steels that have the ability to be made stronger and tougher
because they have good response to heat treating.

4140, 4142, 4147, 4150 – Direct hardening low alloy steels with chrome and
   molybdenum

4340 – Nickel–chrome–molybdenum direct hardening steel used where
   exceptional strength and toughness is needed.

4617, 4620 – Nickel–molybdenum direct hardening steel

6120, 6150 – Chrome–vanadium steels

8617, 8620 – Nickel–chrome-molybdenum steels used for surface (case)
   hardening

52100 – High carbon bearing quality steel with chrome additive

When a piece of equipment is designed the engineer typically looks up a table, like
the one above, to find a material that has the needed mechanical properties, and these
tables can be found in numerous texts and also in manufacturer's literature. Some
points to be kept in mind while reading these tables are:

- Is the value listed the typical or the minimum for that situation?
- Poor heat treatment can create an alloy that is weaker than the strength
  listed for the hot rolled metal.
- Don't hesitate to call a local metallurgist or the vendor's metallurgist for
  advice when choosing an alloy.

UNDERSTANDING STEEL TERMINOLOGY
AND MATERIAL DESIGNATIONS

Remember the cake mix analogy and all the variations that can be made for special occasions. Well, the same thing applies for steels and there are a multitude of alloys available – just like there are many different bread and cake mixes!

Before looking at basic steel and its variations, a review the difference between *wrought* and *cast* products shows:

- Cast materials are poured from a furnace and formed into the finished shape while they are molten. A cast pipe flange involves pouring molten metal into a mold and allowing it to cool. The piece comes out of the mold looking like the finished product, except for some minor machining. The most common cast materials are iron, zinc, steel, and aluminum.
- Wrought materials are made from a material that is molten in the furnace, poured into a shape, then physically deformed into the final shape. For a wrought pipe flange, the molten material is poured from the furnace into a large rectangular billet that is then rolled (hot forged) into a thinner plate in a number of steps. When the desired thickness is reached a smaller piece is cut from that thinner plate and the flange machined out of it.
- Welds are cast materials that are frequently used to join two other materials.

By the nature of the process used to produce them, most cast products have properties that are relatively uniform in all directions, i.e., they are not *isotropic*. On the other hand, with wrought and forged products, because the processing deforms them and results in grains that are stretched in one plane more than the other two planes, they have more strength and toughness in the rolling direction than across the plane of the rolling. Also, because of uniformity and other metallurgical phenomena, there is frequently some advantage to cast products where corrosion is involved and there is generally an advantage to forged products when impact loading is involved.

Hot rolled materials are formed while the basic material is at greatly elevated temperatures and the resultant product does not exhibit much grain elongation or deformation. For steel, this would be in the vicinity of 1500°F (815°C) and for aluminum around 550°F (300°C). *Cold drawn* (also called *cold worked* and *cold rolled*) materials are rolled out while the pieces are at relatively low temperatures, the resultant grain structure is elongated in the rolling direction, and the pieces are harder, tougher, and stronger than if they had been hot rolled. (The technical term for these cold drawn materials is *strain hardened*.)

Q & T stands for quenched and tempered and tells you that the material has been heat treated and is stronger than an annealed or hot rolled product. The heat treating consists of heating to an elevated temperature, rapid cooling, and then reheating to a lesser temperature to relieve the internal stress and both reduce hardness while increasing the toughness.

The concept of tempering is one that is very important to understand and it was one of my early practical metallurgy lessons. I was driving home one day in a standard shift car that had about 84,000 km (52,000) miles on it and it was just out of warranty when there was a series of crunching noises out of the transmission. We towed the car home, jacked it up, and pulled the cluster gear (see Photo 11.1) out of the transmission.

The tops of several the teeth on one of the gears had broken off and I had no idea what caused it, so I asked a friend of mine, a metallurgist at a gear company, what had happened. When I showed him the gear, I knew from the look on his face that he was unhappy. It turned out that his company made the gear and it failed because of a processing problem.

Automotive gears, and most industrial gears, are case hardened and that means the outer layer of the teeth is hardened. When steel is hardened it grows in volume and that creates a stress between the core of the tooth and the case. To reduce the effect of this *residual stress* the gears are tempered. But the gear in our car wasn't properly tempered and the teeth were left with a residual tensile stress. That residual stress, combined with the normal operating stress, was enough to cause an early failure.

Tempering a heat-treated metal reduces the hardness and makes the part tougher and, in doing that, reduces the internal stresses.

T, G & P stands for turned, ground, and polished and usually refers to high quality precision machined shafting material.

CAST IRON

Cast iron is a mixture of iron and carbon with carbon percentage above 1.7%. Common carbon ranges are from about 2.2% to 4.5% and there is a wide range of alloying elements. The result is that it comes in many strength levels and metallurgical structures.

The manufacture of cast iron is an ancient practice and there are many "standards", including gray cast iron, ductile iron, white cast iron, malleable iron, nodular iron,etc. Each of these general groups has different properties resulting from the foundry and heat-treating practices. We've seen huge castings made around the turn of the last century that were as weak as 55 MPa (~8,000 psi) and more modern ones stronger than 700 MPa (~100,000 psi.)

STAINLESS STEELS

There are five different families of stainless steels and their basic divisions and properties are as follows:

- *Austenitic* – These are the familiar 300 series stainless steels. They typically contain large amounts of chromium and nickel for corrosion resistance and some have other additions such as molybdenum for pitting resistance. They are not magnetic and cannot be thermally heat treated but respond well to cold working and are readily weldable. Types 302 and 304 are the most common. (Some heavily cold worked 300 series stainless steels are very slightly magnetic.)

- *Ferritic* – These are some of the 400 series stainlesses and have high chromium contents but little else in the way of alloying elements. They are magnetic, cannot be hardened by heat treatment, and can only be moderately strengthened by cold working. They have good ductility and excellent resistance to stress corrosion cracking. The most common ferritic stainless is Type 430.

- *Martensitic* – These are thermally hardenable, magnetic, straight chromium alloys with moderate corrosion resistance. In the annealed state their ductility and corrosion resistance are comparable to the ferritic alloys. However, their ability to be thermally hardened to tensile strengths greater than 1400 MPa (~200,000 psi) makes them valuable for applications where good corrosion resistance and strength are required. Common alloys include Type 410, which is used for general purpose applications, and Type 440C, which is primarily used for cutlery and industrial knives.

- *Precipitation hardening* – These magnetic alloys typically contain chromium and nickel as their principal alloying elements, but they also have small quantities of columbium, aluminum, copper, and/or tantalum as their precipitating agents. Their primary advantage is that parts can be machined in the annealed state then, at relatively low temperatures, less than 600°C (~1100°F), precipitation hardened to very high strengths with minimal distortion. Their corrosion resistance and ductility are comparable to the similarly hardened martensitic alloys. Common grades include "13.8" (13% chrome and 8% nickel), "15-5", and "17-4".

- *Duplex* – The duplex stainless steels have a structure that is a combination of austenite and ferrite, with a combination of the good corrosion resistance of the 300 series austenitic and the stress corrosion cracking resistance of the 400 series ferritic stainlesses. The result is an alloy that has both good corrosion resistance and high strength. It cannot be thermally hardened and can be welded but temperature control is very important. The most common grade is Alloy 2205 but there are several others plus specialty heat exchanger tube materials.

STRENGTHENING METALS

There are two ways of strengthening metals, heat treating and cold working. Some metals such as the common steels can be strengthened by either method while others, such as austenitic stainless steels, can only be strengthened by cold working. In the next section, we discuss these alternatives and how and why each is used.

HEAT TREATING

When steels are hardened by heat treating the basic process is that the steel is heated to a given temperature, then quenched by cooling it rapidly in air, oil, or water. (The specific quenchant, air, oil, or water, depends on the type and size of the steel and the desired properties.)

In order to get the preferred microstructure the steel should have at least 0.3% carbon. Low carbon steels, such as SAE 1010, 1020, and 8620 don't have this much carbon and they can be hardened to some degree, but to get the optimum metallurgical properties carbon has to be added. The way this is usually done is that they are heated in an oven surrounded by a high carbon atmosphere. The steel gradually absorbs the carbon, then the parts are removed from the furnace and quenched. The result is a hard exterior shell over a much softer core. There are several terms for the process and the more common ones are *case hardening, surface hardening,* and *carburizing.*

(Two comments about these are:

- Sometimes when small quantities of parts are hardened, instead of putting it in an atmospheric controlled oven, the heat treater will coat the piece with a compound that releases carbon as it is heated.
- There are other surface hardening procedures such as nitriding and carbonitriding, which also are used.)

Higher carbon steels, those with more than about 0.3% carbon, can be directly hardened, i, they don't have to absorb additional carbon and don't have to be put in a high carbon atmosphere to be transformed to a much harder material. These directly hardenable alloys are heated and then quenched to obtain the hardness. Most often this is done with alloys where the resulting hardness is essentially the same throughout the part. These are called *through hardened* parts and don't have the sharp delineation between hardened and unhardened areas seen in surface hardened pieces (Figure 3.14).

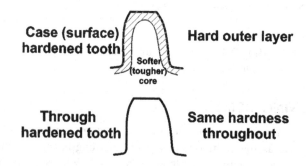

FIGURE 3.14 Comparing case with through hardened gear teeth. In reality, the through hardened tooth may be a little softer on the interior, but the difference is rarely more than a couple of HRC points.

The choice of hardening methods, i.e., surface hardening vs. through hardening, is usually based on production costs but the product usage is also considered. Case hardening is almost always used on things such as automotive gears or roller chains where wear resistance is needed on only a portion of the piece. Through hardening is usually chosen for pieces, such as machine shafts, where increased strength is needed throughout the component. In general, where an entire component is to be hardened, through hardening is less expensive than case hardening.

Some of the other terms involved with heat treating that you will hear are:

Annealed – Heating the part to a specified temperature, around 925°C (1700°F) for steel, then controlled cooling it to reduce the hardness and improve machining.

Normalized – Heating to a temperature above the transformation range, then cooling in air to a temperature well below the transformation range.

Transformation temperature – The temperature, for most steels it is around 730°C (1350°F), above which the grain structure of steel changes from a body centered cubic to a face-centered cubic.

Quenched – After a part is heated up to a given temperature it is rapidly cooled. This is called "quenching" and different quenchants are used such as oil and water, or even air, to obtain different material properties.

Stress relieved – Heating a steel to a temperature to allow for internal reorientation of the structure to reduce stresses that usually result from processing. Most steels are heated to 620 to 650°C (1150 to 1200°F) for one hour for every 50 mm (one hour per inch) of thickness, then slowly cooled. (Stainless steels have to be stress relieved at approximately 1000°C [~1800°F] to avoid sensitivity to corrosion problems.)

Carburized, carbo-nitriding, nitriding, and cyaniding – All refer to "surface hardening" methods that are usually done in an oven and result in a thin hardened layer over a softer core. Carburizing involves carbon. Nitriding and cyaniding both use nitrogen compounds. Carburizing usually gives a thicker case but is more expensive.

Through hardening – Refers to a part that is essentially the same hardness throughout as compared to surface or case hardened pieces.

Flame and induction hardening – Are surface hardening methods that don't require an oven and don't heat the entire piece. Flame hardening is frequently used in low quantity processes but can be used in higher quantity production machinery. Induction hardening is usually a relatively high quantity production process.

Tempering – Is the procedure where a part is reheated to a lower temperature immediately after hardening. The primary reason for tempering is to reduce internal stresses and add toughness to the material. In case hardened products, tempering results in a slight decrease in hardness with a substantial reduction of the stresses across the interface between the case and the core. Some tempering examples are:

- Gears where the part is originally quenched to HRC 67 or so, then tempered at 275°C (~525°F) for an hour with the result a tooth hardness of HRC 55.
- Bearings where the original hardness is HRC 67 and the part is tempered at 220°C (~425° F) with HRC 61 +/- the result.

COLD WORKING

As mentioned above, a second way of strengthening some metals is by cold working. This involves physically deforming the grains at temperatures well below their transformation temperature and results in an elongated, strain hardened grain structure.

Austenitic stainless steels are particularly good examples because they can't be thermally hardened, but cold working will result in tremendous increases in the yield and tensile strengths. For example, 304 stainless has a yield strength of about 210 MPa (~30,000 psi) in its annealed condition while fully cold worked it can reach as high as 1200 MPa (~170,000 psi).

Some comments:

- *Cold working, cold drawing, work hardening,* and *strain hardening* mean the same thing and refer to the same processes.
- Most of the materials that respond well to work hardening cannot be heat treated to increase their tensile strengths. Two excellent examples of this are the 300 series stainlesses (i.e., austenitic stainlesses) and copper alloys.
- In order for a metal to be cold worked the crystal structure has to be deformed.

Carbon steels can be cold worked, as can be seen by the "Cold Drawn" data in Table 3.4 above. The increase in tensile strengths can range from a minimum of about 5% up to a practical maximum of about 50%.

THERMAL EXPANSION

If we heat any material, solid, liquid, or gas, it will expand. This is because there is increased molecular activity. Some materials, like gases, expand exactly in proportion to the temperature change (referred to absolute 0°). However, with steels and other rigid materials there are *coefficients of expansion* that can be used to determine how much a material will change in any dimension at a given temperature. For instance, around room temperature, low carbon steels grow (or shrink) by 0.0000011 mm for each mm of length for each degree Celsius (0.0000063 in for every inch of length for every °F) temperature change. So, if there is a 500 mm (19.7in) long bar at room temperature and it's heated to 225°C (437°F), the change in length (ΔL) can be found by:

$$\Delta L = \varepsilon L \Delta t$$

Where: the change in length is ΔL, the length is L, the change in temperature is Δt, and the coefficient of expansion is ε. Therefore, if the bar starts out at an ambient temperature of 20°C (68°F), the change in length is:

$$\text{metric } \Delta L = 0.0000011 \, mm/mm/°C \times 500 \, mm \times 205°C = 0.11 \, mm$$

$$\text{U.S. } \Delta L = 0.0000063 \, in/in/°F \times 19.7 \, in \times 369°F = 0.046 \, in$$

Another example that gives a substantial insight into the deformation and residual stresses caused by welding is to calculate the amount of contraction that occurs in steel weld between the point where the weld puddle solidifies and room temperature. An alloy such a steel really doesn't freeze at a single temperature like water, but let's assign the "freezing point" as 1500°C (~2700°F), then calculate the shrinkage in a 61 cm (24″) long weld as the part cools to room temperature. With the length as 61 cm (24 in.), the change in temperature as 1480°C (2632°F), and the coefficient of expansion 0.000011 cm/cm/°C (0.0000063 in/in/°F), $\Delta L = \varepsilon L \Delta t =$

$$\text{metric } \Delta L = 0.00000011 \, cm/cm/°C \times 61 \, cm \times 1405°C = 0.99 \, cm$$

$$\text{U.S. } \Delta L = 0.0000063 \, in/in/°F \times 24 \, in \times 2632°F = 0.40 \, in$$

(In reality the coefficients of expansion of many metals change somewhat with temperatures.)

Table 3.5 shows various expansion coefficients and from it we can see that if a part made from:

- Aluminum – The expansion would be twice that of carbon steel.
- 300 series stainless steel – The expansion would be about 1.2 times as much as steel.
- Plastic – The expansion could be ten times as much as steel!

There are many generic or slang terms used for metals in industry. Some of the more common ones are:

- Black iron – as in "Black Iron Pipe" – is ordinary low carbon mild steel pipe usually with cast ductile iron fittings.
- White metal was originally a term used to describe any of several tin, lead, and antimony alloys used for babbitt bearings. Now frequently used to refer to any cast aluminum, zinc, and magnesium alloys.
- A 36 plate – Refers to ASTM A 36 steel plate. A low carbon good quality mild steel with a minimum yield strength of 36,000 psi. (250 MPa)
- Wrought iron – Many years ago it was a very low carbon steel that was not very strong but easily worked. Today it is a generic term for mild steel.
- Cast steels can have basically the same properties as wrought steels with the same chemistry.
- Low alloy steels – Steels with less than 6% alloying elements
- High alloy steels – Stainless and other specialty steels generally with more than 10% alloying elements.

TABLE 3.5

Coefficients of Expansion for Various Materials

Material at Room Temperature	Expansion Coefficients	
	mm/mm/°C	in/in/°F
Steel	0.0000117	0.0000063
Aluminum	0.000023	0.000013
Brasses (range)	0.000018–0.000022	0.0000096–0.000012
Brick (typical)	0.0000047	0.0000026
Cast iron	0.0000105	0.0000058
Copper	0.0000165	0.0000089
Glass (range)	0.0000063–0.0000090	0.0000035–0.000005
Lead	0.000029	0.0000016
Nickel	0.0000133	0.0000072
Plastics (range)	0.000064–0.000198	0.000036–0.00011
Stainless steel – austenitic	0.000017	0.000093
Stainless steel – martensitic	0.000013	0.000070
Stone (range)	0.0000079–0.000012	0.0000031–0.0000079
Titanium	0.000095	0.000053
Zinc	0.000034	0.000019

Notes:

- Wind Chill Factor *does not* apply to materials. It only applies where there is evaporating moisture, such as human skin or water-wetted components.
- The coefficients apply in the temperature range between –20°C and 150°C (0°F and 300°F). For other temperature ranges there are other coefficients. For instance, when steel is heated from 700°C to 750°C (1300°F to 1400°F) it actually shrinks!
- These are typical values and there will likely be variations between alloy groups, and so on.

Temperature Effect on Tensile, Fatigue and Yield Strengths

Earlier in this chapter we mentioned that there is a tremendous change in toughness (impact resistance) of many steels between about –25°C and 100°C (–10°F and 212°F) but there isn't much change in the other strengths until they become a lot hotter. In the next chapter Figure 4.2 is a graph that shows the short-term effect of elevated temperatures on the tensile strength of several alloys.

Chapter 3 Summary

A basic understanding of metallurgy is critical to effective failure analysis.

All steels deflect the same amount with a given load. The advantage of going to stronger materials is that they deflect more before permanently deforming. The downside of these stronger materials is that there is a narrower window between the yield strength and tensile strength.

When a part fails, the fracture plane is always perpendicular to the plane of maximum stress.

Looking at the different loading mechanisms, impact loads, steady state loads, and fatigue loads all rely on different material properties.

The major benefit of alloying elements is that they improve the toughness and hardenability of the parent metals.

BIBLIOGRAPHY

Handbook of Chemistry and Physics, CRC Press, Boca Raton, FL, 2006.

Metals Handbook, Volume 1, Properties and Selection, American Society for Metals, Metals Park, OH, 1990, ISBN: 978-0-87170-377-4.

Metals Handbook, Volume 8, Failure Analysis, American Society for Metals, Metals Park, OH, 1979, ISBN: 978-0-87170-389-7.

Nickel Development Institute, *Design Guidelines for the Selection and use of Stainless Steels*, www.nickelinstitute.org/media/1667/designguidelinesfortheselectionanduseofstainlesssteels_9014_.pdf.

Three Keys to Satisfaction, Climax Molybdenum Company, New York, 1961.

United States Steel Corporation, *Suiting the Heat Treatment to the Job*, United States Steel Corporation, Pittsburgh, PA, 1958.

4 Overload Failures

There are different basic systems for classifying failures and, as outlined in Chapter 2, we've divided them into four categories, overload, fatigue, wear, and corrosion. We'll start by reviewing the major features of each of these categories, then investigate them in detail as the failure diagnoses are applied to specific parts. This chapter will deal with overload failures.

INTRODUCTION

Overload failures essentially happen instantaneously. When a glass bottle falls to the floor and shatters, the tensile strength has been exceeded. When that happens, a *brittle fracture* occurs and the cracks grow with incredible speed, over thousands of meters (or feet)/second. However, with a *ductile overload failure*, such as the collapse of an inexpensive wire hanger or even a metal building roof, the metal bends and buckles and fails, not as rapidly as a brittle fracture, but much, much more rapidly than fatigue, wear, or corrosion failures.

We have divided overload failures into two classifications – ductile failures and brittle failures – based on the differences in their appearance. For example, if identical overload forces act on two pieces, one a ductile material (mild steel) and the other a brittle material (ceramic), the ductile piece will show easily visible deformation, while the brittle part breaks into pieces. Furthermore:

- Once the ductile piece is deformed, it can no longer perform the design duties, however if the stress persists the part may continue to stretch and eventually fracture. (As mentioned in Chapter 3, the stress that causes it to deform exceeds the yield strength while the stress at the point when it finally fractures is the tensile strength or greater.)
- With a brittle fracture the yield and tensile strengths are almost identical, so there is effectively no warning of a failure, very little or no measurable elongation, and the part suddenly breaks into pieces when the load becomes great enough.

Glass is an example of a brittle material and marshmallows are an example of a ductile one, but not many materials are this obvious. There will be some industrial materials that seem to be half-way in between and looking at some of the common materials used in industry, they can be divided as follows:

- *Ductile materials* include mild (unhardened) steel, most aluminum alloys, most copper alloys, titanium, austenitic stainless steel, silver, tin, many plastics, etc.
- *Brittle materials* are ceramics, cast iron, graphite, concrete, hardened steels, glass, some cast aluminum alloys, wood, some plastics …

Metallurgists generally define brittle materials as those with less than 15% elongation between the yield point and final fracture.

When looking at overload failures it is important to recognize the differences in appearances between the brittle and ductile failure modes and Figure 4.1 shows how, with the same loads, the appearances are very different.

An example of that difference in appearance can be seen in the identically loaded parts shown in Photo 4.1. These are three chain rollers used in industrial bucket

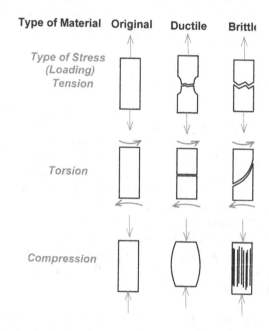

FIGURE 4.1 Showing how the different forces and different material properties result in very different overload failure appearances.

PHOTO 4.1 Three conveyor rollers. On the left is a case hardened piece with a series of brittle fractures. The center one is unused, and the unhardened one on the right shows the effect of a ductile overload.

elevators and they function very much like the rollers in a common roller chain. The center piece is as they were supplied and the pieces on either side were from different production batches but used in the same application. (They both should have been case hardened with a relatively thick case, but the supplier made an error.) The one on the left was hardened and has fractured longitudinally in several places while the unhardened one to the right has tremendous ductile deformation. Identical compressive loads with very different failure appearances because one was brittle and the other ductile.

Photo 4.2 is an example showing two torsional overload failures and the very obvious difference between ductile and brittle materials. The piece on the upper left is from a hardened utility truck axle, and this brittle fracture occurred only a few thousand miles after the truck was delivered. The cracking started at the large metallurgical defect shown by the arrow and then propagated at a 45° angle to the central axis. Below it and to the right are two pieces from the ductile overload failure of an inexpensive 1/2″ socket wrench extension and the end has been smoothly turned off.

Some important points that can be realized from overload failures are:

- They very clearly show the part was tremendously overloaded.
- The stress that caused the failure was applied immediately before or as the failure occurred.
- The direction of the fracture describes the direction of the failure load.
  - With brittle fractures the crack is always perpendicular to the failure force.
  - With ductile failures the direction of the elongation is the direction of the failure force.

(The initial reaction may be that the last statement seems contradictory. But what happens when a brittle piece is compressed is that it tries to grow outward in the shape of a barrel. However, the lack of ductility results in the hoop stresses causing multiple cracks.)

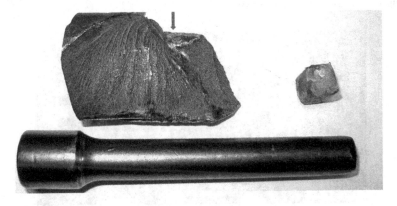

PHOTO 4.2 Two torsional overload failures with the brittle fracture in the upper left and the two pieces of a ductile overload failure.

So, if two identical appearing parts, one of an unhardened mild steel and the other one of a hardened steel, are overloaded identically, the ductile mild steel part will bend, while the brittle hardened part cracks or snaps in two. The failure analyst can look at both of these parts and rapidly see:

- That they were the result of loads applied shortly before failure.
- The direction of the forces that caused the failures.

The spline shown in Photo 4.3 is a bit of an uncommon example because it didn't occur as the result of a single blow. The yield strength of a metal is generally defined as occurring when the plastic deformation exceeds 0.2%. But this shows what eventually happens when a load that causes a 0.01% or a 0.005% deformation is applied many, many times.

Another example of that repeated very slight *ductile overload* is shown in Photo 4.4. This is a coupling piece from a set of steel mill rolls where the loads have been repeatedly increased in an effort to keep up with competitive financial pressures. It started out with straight flutes but, as can be seen in the photo, the coupling steel is very ductile!

On a microscopic basis there is a huge difference between the ductile and brittle failure mechanisms. Ductile failures occur through rupture of the individual grains of metal and generally require a great deal of energy to propagate the crack. On the other hand, brittle fractures grow by separating the grains along their boundaries and usually don't take much energy to propagate. Occasionally, in the brittle fracture of a ductile material such as the motor shaft in Photo 4.12, the crack will stop growing because it just doesn't have enough energy to continue tearing the metallurgical structure apart.

TEMPERATURE EFFECTS ON OVERLOAD FAILURES

For most metals their ductility, or lack of ductility, is an inherent property and changes very slowly as temperatures increase. Figure 4.2 shows how there is really not much effect in yield strength until about 300°C (600°F) and even above that the strength decreases relatively slowly.

PHOTO 4.3 The ductile deformation of the input spline on a small reducer.

PHOTO 4.4   A seriously twisted steel mill coupling. It was deformed by repeated torsional loads. (To give you an idea of the scale, that's a 75 mm [3″] wide strap wrapped around the coupling and a sledgehammer handle just to the right of the strap.)

However, that doesn't apply to the impact strength and Figure 3.7 shows the tremendous effect of temperature changes on the toughness of four common steel alloys. (Ni-Mo PV is a nickel-molybdenum pressure vessel steel.) So, even though there's little difference in tensile, yield, and fatigue strengths between about –75°C and 300°C (–100°F and 575°F) over that same range of temperatures, impact strengths can vary by more than a factor of five. This loss of impact strength as the temperature decreases is usually called *low temperature embrittlement*.

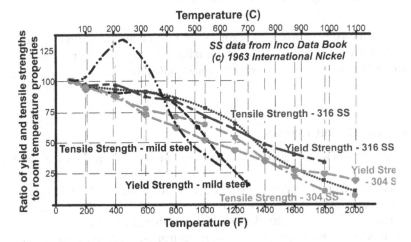

FIGURE 4.2   The temperature effect on tensile and yield strengths of mild steel and two stainless steel alloys.

An interesting example of this low temperature embrittlement phenomenon can be seen in the failure of a group of chairs from ski lifts. Ski lifts most commonly operate from about −18°C to 2°C (0°F to 36°F) and these chairs were impact loaded when they swung and hit the top terminal support. Several of them failed even though they were empty at the time of failure and, like all failures, there were multiple causes. But one of the contributing roots was that they were made from an SAE1030 carbon steel instead of the specified SAE 1020. The 1030 was about 25% stronger than the 1020, yet it was the stronger chairs that were failing! ... And, by looking at both the fracture faces and Figure 3.7, we can understand why.

On that ski lift, the impact loads occurred as the chairs swung and hit the support for the wire rope wheel at the top of the ski hill. At the operating temperatures the specified weaker material would require between 100 and 115 joules (75 and 85 ft-lbs.) of energy while the stronger SAE 1030 chairs were falling because their impact strength was about one-fifth of that.

The first time we got involved with a low temperature embrittlement failure was in an ancient manufacturing plant. Many years earlier, someone had fabricated a rigging bracket from a piece of SAE 1045 steel and had welded it to a column. For years the bracket was used every summer for lifting up to 2000 lbs. of material to the roof of a building. Then one year the outage was held on a very cold day in November and the bracket snapped when there was almost no load on it. There is always a trade-off for higher strength!

An important point to recognize is that this low temperature embrittlement does not affect the common 300 series stainless steels. In that range shown in Figure 3.7 where carbon steel toughness drops precipitously, a typical austenitic 304 or 316 stainless part has a Charpy V-notch toughness well over 135 joules (100 ft-lbs.) that actually increases slightly as the temperature decreases.

One additional point to repeat – Almost everything is designed with a safety factor and the existence of an overload failure is proof that the part has been stressed far beyond what the original design called for.

ANALYSIS OF DUCTILE FAILURES

Some key points in looking at a ductile failure and understanding the forces involved are:

- The fact that an overload failure happens proves that the load was greater than the yield strength of the part.
- The distortion lies in the plane of the forces that caused the failure.
- The applied final fracture force was greater than any of the previous loadings.
- Microscopically the failure happens when the metal grains are torn apart.
- Ductile fractures usually require a great deal of energy.

Below is a series of photos that illustrate some ductile overload failures (Photos 4.5–4.7).

The two parts of Photo 4.8 (A and B) show evidence of the elastic overload of a weldment on a piece of construction equipment and they show the ends of the reinforcements that project through both sides of a box beam on a crane. The cracked paint clearly shows the stress fields around the ends of the welds. Notice that the left photo shows much more cracking than the other. The paint cracks are proof that this symmetrical beam was side loaded and much more in one direction.

PHOTO 4.5   This shows the ductile torsional overload failure of a splined reducer shaft. Note the twisting (ductile) distortion of the splines. (This shaft failed when another piece failed and jammed the reducer.)

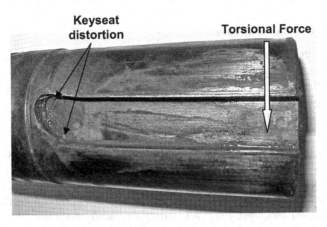

PHOTO 4.6   shows the ductile distortion of the end of a mild steel motor shaft and the end of the keyway has been seriously deformed. The distortion is at the end of the keyway because the stress was greater than the yield strength of the steel, but the portion of the shaft under the coupling was stronger because it was reinforced by the key and coupling hub fit.

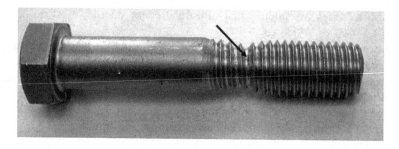

PHOTO 4.7   is a ductile tension overload of a 16 mm (5/8″) bolt. The arrow points to the beginnings of the classic cup and cone fracture usually seen in the later stages of a pure tension failure.

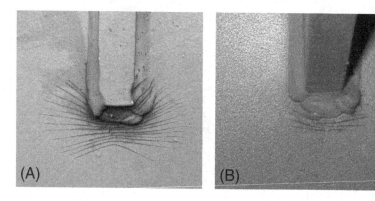

PHOTO 4.8   (A and B) These are two views of the reinforcements on the side of a hydraulic crane beam. The cracks in the paint show this crane beam was both overloaded and side loaded, and it's pretty obvious that the load wasn't centered!

ANALYSIS OF BRITTLE FRACTURES

Some of the key points in a brittle fracture are the same as a ductile failure, i.e., the load was greater than the yield strength of the part, the fracture force was greater than any of the previous loadings, and the load was applied immediately before the failure took place.

Some noteworthy differences are:

- In a brittle fracture, the fracture plane is perpendicular to the direction of the forces that caused the failure.
- A brittle fracture uses much less energy than a ductile failure. Microscopically the brittle failure happens as a fracture between the grains (along the grain boundaries) and, once the crack starts, it typically does not take a lot of energy to propagate.
- The point where the failure started can usually be seen by following the "chevron marks", lines on the surface of the fracture face, that point toward the origin.

Photo 4.9 shows the brittle fracture of a large reducer input shaft that was caused by an impact load when the shaft was bumped. The "chevron marks", the lines on the fracture surface, are a great aid in the diagnosis of brittle fractures because they point to where the failure started. An interesting point is that most persons would expect the fracture to begin at the keyway notch, but the actual origin is just to the left of it.

In Photo 4.10 can be seen the brittle tension fractures of two pulverizer rolls. Excessive hoop stress caused the fractures of these extremely hard and strong but very brittle castings. The rolls had been shrink-fitted onto shafts and the cracks run straight down the side of the roll, perpendicular to the hoop stress. The rolls show no visible distortion or elongation and the cracking is proof that the shrink fit was excessive.

Chevron Marks

As shown in several of the photos above, a valuable feature on the face of a brittle fracture is that the "chevron marks" point to where the failure started. They show

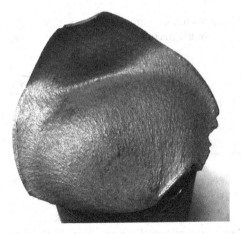

PHOTO 4.9   The brittle fracture of a reducer shaft with "chevron marks", lines on the surface of the fracture, pointing toward the crack origin.

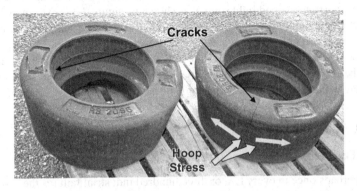

PHOTO 4.10   Two pulverizer rolls that have brittle fractures, i.e., cracked from hoop stress because they were too tight on their mounting shafts.

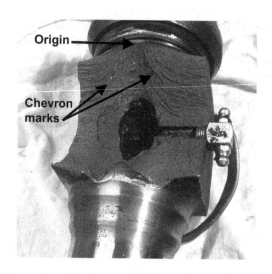

PHOTO 4.11 The brittle fracture of a universal joint with very obvious "chevron marks" that point to the origin. The brittle fracture also proves that the operating load was excessive.

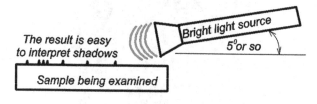

FIGURE 4.3 This shows how you should use the light to accentuate the surface features of any fracture and aid in the interpretation.

where to take the metallurgical samples needed to check as to whether the metallurgy or condition of the material played a part in starting the failure.

There are times when it is difficult to see the chevron marks and a careful inspection with a narrow beamed light source, as shown in Figure 4.3 is extremely helpful. Experience has shown that the best interpretation of surface features can be done with a narrow, bright light beam at a very shallow angle.

Photos 4.12A and B demonstrate the actual effect of having that light at a low angle. These are two pictures of the brittle fracture face of a crane hook and the *only* thing changed between them was the angle of the light beam.

UNUSUAL CONDITIONS

BRITTLE FRACTURES OF DUCTILE MATERIALS

Rarely, perhaps once in every two or three hundred industrial failures there will be a brittle fracture of a ductile material. Every history buff knows the story of the Liberty ships in World War II and their problems with brittle fractures. Several of these

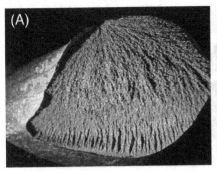

PHOTOS 4.12 (A and B) Two views of the same fracture with the only difference being the angle of the lighting. Note how clearly the "chevron marks" show up and point to the fracture origin in the left photo. (The cracking started on the side of the hook and proved that the hook was side loaded.)

ships were lost in the North Atlantic with their destruction initially being blamed on German submarines. Even though the materials of construction were ductile, many of these Liberty ships suffered major fractures essentially without warning and one even sank in the harbor before being loaded. This outbreak of brittle fractures resulted in several detailed studies of the causes of brittle fracture and a much better understanding of the problem.

Brittle fractures of ductile materials essentially take place under one of the following conditions:

1. The part is loaded so highly and so rapidly that the ductile material doesn't have time to deform before it is fractured.
2. The ductile material is so constrained that it can't deform before it fractures.

Either the ductile materials don't have enough time to behave in a ductile manner or the design of the structure is such that the material can't deform in a ductile manner and the result is that a brittle fracture occurs.

(Don't confuse this with the *low temperature embrittlement* of ductile materials. In low temperature embrittlement the piece fails because it is impact loaded but doesn't have the toughness to resist catastrophic failure. In these failures, the result is the same, but the parts are not impact-loaded.)

Rapid Force Application

Two examples of brittle fractures of ductile materials that were caused by the rapid application of a force are:

1. In one of the analyses we worked on, a cable-supported industrial elevator mechanism fell about 30 m (100 feet). When the carriage reached the end of the cable, traveling at a high speed, the cable snapped like a piece of string and, because the load was applied so rapidly, the ends of the ductile wires looked like brittle fractures.

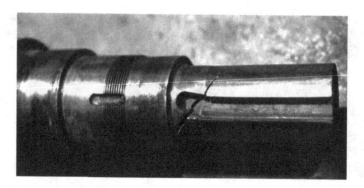

PHOTO 4.13   The end of a very ductile 3600 rpm motor shaft that shaft that stopped so fast the fracture might be mistaken for a brittle overload.

2. Photo 4.13 shows the coupling end of a 150 kW (200 hp), 3600 rpm mixer motor shaft with a diagonal crack in it. The mixer motor was running at full speed when some product was accidentally dropped into the machine, immediately jamming and stopping the rotation. An analysis showed the shaft was SAE 1035 steel, a relatively ductile material that, according to Table 3.4, should elongate by about 18% before failure. Nevertheless, one can easily see a diagonal crack that appears to be the torsional failure of a brittle material. However, further examination of the parts and the site shows:

    a. The fracture is not on a 45° angle as would be expected from a purely torsional brittle fracture.

    b. There is some deformation at the end of the keyway.

    c. The crack stopped half way across the shaft.

Looking at these apparent incongruities, points a and b happened because there was *some* ductility to the shaft and it had enough time to deflect a little, but not what would have normally been expected. (If the shaft had reacted in a purely ductile manner the end of it would have been turned off and the fracture would have been essentially straight across the shaft, like the inexpensive socket extension in Photo 4.2.) Point c shows that there wasn't enough energy to drive the crack all the way across the shaft because the failure was converting from cleavage to ductile rupture.

Constrained Materials

The other situation where ductile materials act in a brittle manner is when they are so constrained that they can't deform. (This was the basic cause of the Liberty Ship failures.) For example, if there is a square corner weld where several heavy plates are welded together with significant residual stress, the weld material cannot deform to release the stress in a ductile manner because it is being pulled in three directions at once. The result is a fracture, which often occurs at stresses below the yield point.

One example of this type of failure happened with the welded structure of a process tank. There were several 10 mm and 12.5 mm (3/8″ and 1/2″) sections welded together as a reinforcement and the reinforcing steel was overstresed. Immediately next to the joint another piece was welded in. The 12.5 mm (1/2″) thick section necked down (ductile deformation) about 20%, then snapped in a brittle manner. Along about 10 cm (4″) of the bar, the crack looked like ductile elongation but the remaining 35 cm (14″) fractured as a brittle overload.

Photo 4.14 shows the brittle fracture of another highly constrained weld. (The small black spheres in the photo are slag from the process operations.) The photo shows the fractured weld on a lifting beam in a steel mill and where a 6 mm (1/4″) thick heat steel shield is attached to a 25 mm (1″) thick beam. The weld material has a catalog elongation of about 25%, yet it fractured in a brittle manner. This happened because the thinner piece heated and expanded much more rapidly than the heavy section, so the weld was very heavily stressed. It couldn't deform in a ductile manner, because it was a 6 mm (1/4″) throat tack weld on a 25 mm (1″) thick plate, so the weld snapped!

An incredibly loud and impressive example of a brittle fracture involved a large 50-year-old (1.5 m, 4.5 ft. diameter) heat exchanger shell in an office building. The 25 mm (1 In) thick vessel shell was made from ASTM A 36, a very ductile material, but very shortly after a substantial modification there was a noise that was described as "sounding like someone had blown off a bomb in the building". A contractor had modified the shell by welding in a long section of 200 mm (8 in) pipe, but didn't put any supports under it, so it put a huge bending load on the shell. Not long after the installation there was a noise like an explosion and the inspection found a 900 mm (36 in) rupture of the vessel wall that also extended well into the pipe. A great example of a brittle fracture because the heavily stressed material was so constrained it couldn't deform.

Notch Sensitivity of Brittle Materials

In Chapter 5 on fatigue failures there is a lengthy discussion on stress concentrations (sometimes called stress risers), how they increase the stress in their immediate vicinity, and how they contribute to fatigue failures. In a similar manner, brittle materials

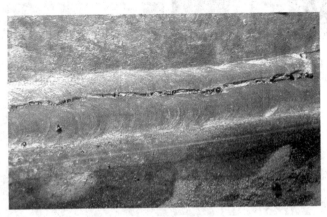

PHOTO 4.14 The brittle fracture of a ductile weld material because it was so constrained that ductile elongation wasn't possible.

are very susceptible to sharp corners and notches. If the notch is severe enough and the stress is high enough, the part will fracture even though the field stress may be less than 75% of the yield (or tensile) strength. Below we show two examples of this.

Photos 4.15 (A and B) – These show two views of a cast alloy hammer from a 1100 m ton/hour stone crusher. This very hard material iron alloy is designed to be abrasion-resistant, but it is relatively notch-sensitive, with the notches highlighted by the arrows in both photos.

Photo 4.15A shows the new hammer assembly after it has shattered, pretty much destroying everything in sight. The second photo is a close-up of the left side of the hammer showing an obvious brittle fracture with the chevron marks pointing toward two small black areas. (They are black because they have oxidized over time.) These casting defects resulted in the notches that substantially weakened the part.

In the example above, the load was applied very rapidly but for a notch to cause these problems doesn't require impact loading. Photo 4.16 shows a fork from a 36 m ton (40 ton) capacity lift truck that was slowly carrying a heavy load across some uneven ground when the fork snapped. The notches that caused the failure can readily be seen and it is obvious from their dark color that they had been there for some time. Some comments are:

- The photo was taken in the field and it had been raining.
- The load was much less than what the forks were rated for and was properly positioned.
- The notch was a stress concentration and the hardened heat-treated material has poor notch sensitivity.

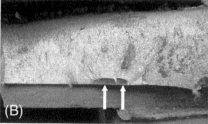

PHOTOS 4.15 (A and B) Two photos of the brittle fracture of a hardened cast alloy crusher hammer. The notches, shown by the arrows in both photos, were from a manufacturing defect.

PHOTO 4.16 The brittle fracture of a fork off of a lift truck with an older cracked area in the top center that created a stress concentration.

THREE VALUABLE BRITTLE FRACTURE EXAMPLES

A Case Hardened Bell Crank

The photos show a 130 mm (5″) diameter pin used as a pivot in the boom of a very large bucket loader used in a production operation. The pin had an HRC 60 hardened outer layer about 3 mm (1/8″) thick while the core was about HRC 20. Photo 4.17 shows the origin while the lighting has been changed in Photo 4.18 to emphasize several earlier cracks. Inspecting the fracture face shows that:

- The hardened case fracture surface is much smoother than the core (because it has a finer grain structure).
- The early smaller cracks were from impacts that didn't have enough energy to propagate through the entire case.
- Using low magnification, about 7X, and low angle lighting, the chevron marks in the case can be easily seen.
- The propagation marks continue through the unhardened core but are much coarser.

PHOTO 4.17   A close up showing the chevron marks in the hardened case.

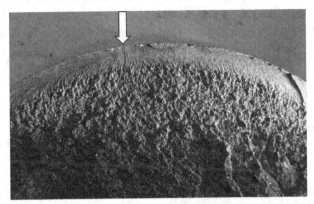

PHOTO 4.18   The same piece with a little different lighting. The origin is shown by the arrow and the semicircular earlier cracks can be seen just to the right of the origin.

PHOTO 4.19   Two very ductile stainless steel bolts that were grossly overloaded and fractured in a brittle manner.

From the existence of those small earlier cracks we can tell that the impact load that caused the failure wasn't an isolated event. (Further investigation found that the bucket loader operator occasionally ran into some of the building structure!)

Brittle Fracture of Two Very Ductile Stainless Bolts

Photo 4.19 shows two low strength Type 316 stainless steel bolts from a paper machine. They failed immediately when the machine was started up after some maintenance. After a brief inspection we found that some tools had been left where they smashed against the bolted bracket as the machine started. ASTM A 593 C bolts should be very ductile, but looking at them in the photo, the one on the left shows obvious chevron marks and no deformation while the one on the right shows some deformation, but far less than a normal ductile overload. From that we can tell that the two were instantly tremendously overstressed.

A Great Welding Metallurgy/Brittle Fracture/Failure Analysis Example

In Chapter 3 we talked about how hardening of a metal increases the volume and we also wrote about how it is important to understand the chemistry of a steel alloy before trying to weld it.

One of the things that we've found time and time again is that people will try to improve or fix something but not understand the effects of what they do. I like it that people try to improve things, but wish that they would stop and think a bit about:

1. Why their problem is happening
2. Do some research, maybe just look on the internet or in some sort of guide book, before they jump in and make changes

The people responsible for the next example had no clue as to what damage they could cause with their modification. But the saddest thing is that, if they had asked why the parts were failing, they could have easily and inexpensively fixed it.

The plant had a bucket elevator with a large engineered chain that the buckets were attached to. The chain links were held together with cotter keys in the chain pins, but the problem was that the links kept shearing the cotter keys and then the chain would come apart.

(At this point they should have asked, "How can there be a side load on a cotter key?" because those keys are designed to just keep pieces from falling apart. But they didn't do that and didn't recognize that there was a side load on the chain caused by the headshaft misalignment.)

So they decided to weld some cross supports between the side links to prevent them from coming apart. Photo 4.20 shows a broken link with the battered welded cross support. Photo 4.21 shows a view of the fractured link and looking at it we can clearly see the chevron marks that point to the origin and the hardened nugget at the toe of the weld.

What happened was that the side link was SAE 1045 steel and, with 0.45% carbon, was easily hardened. But they didn't preheat the side link before welding and the mass of the part quenched the toe of the weld creating a hardened area that had expanded in volume, creating an internal stress in the part. Then, when the abnormal operating stresses were applied the link snapped.

PHOTOS 4.20 AND 4.21   On the left is the broken chain with the welded cross support that the plant had added. (Note that the weld on the end of the support has been deformed in a ductile manner.) On the right is the brittle fracture of the side link with obvious chevron marks pointing to the origin at the weld toe.

CHAPTER 4 SUMMARY

Overloading a part can result in either a ductile or a brittle fracture depending on the material characteristics and the rate of load application.

Most of the materials we deal with on a regular basis are ductile.

There is a tremendous difference between tensile strength and impact strength. The impact strength, i.e., toughness, for many steels change significantly between −15°C and 50°C (0°F and 125°F) while the tensile and yield strengths are essentially unchanged.

Under certain conditions, i.e., when loads are applied very rapidly or the piece is very highly constrained, ductile materials can behave as though they were brittle. (But brittle materials only behave in a ductile manner at highly elevated temperatures.)

Generally, the harder and stronger a material is, the more sensitive it is to notches.

BIBLIOGRAPHY

Metals Handbook, Volume 1, Properties and Selection, American Society for Metals, Metals Park, OH, 1990, ISBN: 978-0-87170-377-4.

Sachs, Neville W., *Failure Analysis Made Simple: Bearings and Gears*, Reliabilityweb.com, Ft. Myers, FL, 2015, ISBN: 978-1-941872-30-7.

Sachs, Neville W., *Failure Analysis Made Simple: Shafts and Fasteners*, Reliabilityweb.com, Ft. Myers, FL, 2018, ISBN: 978-1-941872-81-9.

Wulpi, Donald, *Understanding How Components Fail*, ASM, Metals Park, OH, 1986, ISBN: 0-87170-189-8.

5 Fatigue Failures (Part 1): The Basics

Fatigue causes more mechanical failure analyses than any other mechanism. It caused the bearings in our camper wheels to rumble and cracked the fan blades on our old car. It varies with the type of machinery, but in most cases, it causes at least 80% and possibly 90% of all mechanical failures. (It is uncommon that we see a fastener failure that doesn't have fatigue as a major contributor.)

By definition, overload failures generally take place during one load application and at loads equal to or greater than the yield strength of a material. On the other hand, fatigue failures generally occur at loads less than that needed to cause plastic deformation and they usually require many stress cycles to initiate and then grow across a part.

When an overload failure occurs, inspection of the failure will tell the direction of the applied force, the rate at which the force was applied, and the fact that the load that caused the failure was applied in the millisecond or so before the part failed. In contrast, the event that started a high cycle fatigue failure may have initially been applied one hundred thousand, a million, or even twenty million or more stress cycles in the past. Fortunately, careful inspection of the part and particularly the fracture face can show the relative load, the direction and uniformity of the load application, whether stress concentrations played an important part, and many other important facts about the contributing causes.

FATIGUE FAILURE CATEGORIES

Fatigue failures are caused by cyclical loads. There aren't any formal definitions for the fatigue failures categories, but three general classifications are:

1. *High cycle fatigue (HCF)* – Where the failure takes more than 10,000 cycles. These probably account for more than 90% of all fatigue failures. They are more common than either of the other categories and this chapter will cover them in detail. Most high cycle fatigue failures take more than 100,000 cycles and some may take hundreds of million cycles from the initial force application to the final failure.
2. *Low cycle fatigue (LCF)* – When the failure takes fewer than 10,000 but more than about 25 cycles from the initial application of the force to final failure, it is considered a low cycle fatigue failure. They are generally not very common in industrial equipment and, referring to the S-N curve in Figure 5.1, one can see they involve significantly higher stresses than high cycle failures.

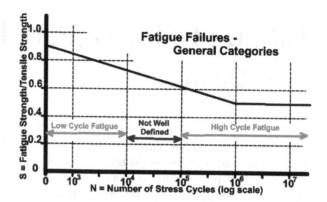

FIGURE 5.1 A Stress-Number of cycles (S–N) curve showing the generally accepted ranges of fatigue stress cycles.

3. *Very low cycle fatigue (VLC)* – One interesting point about low and high cycle fatigue failures is that, after the part fractures and the pieces are examined, they show no visible plastic deformation. (There may be some deformation of the last thread on a failed bolt but there is no visible deformation of the body of the part.) However, VLC failures:
   a. Take only a few stress applications, typically less than 10, from initiation to final failure.
   b. Show some small visible deformation in ductile materials.
   c. Indicate that the applied force was only slightly less than what would have caused an overload failure.
   d. Are not very common in industrial equipment but are seen more often than low cycle fatigue examples.

To the best of my knowledge, fatigue failures cannot occur from pure compressive stresses. For fatigue to happen, the part *has* to undergo periodic tensile stress then relaxation, then stress, then relaxation, and this cycle *has* to be repeated many times. How long does it actually take for a fatigue crack to grow across a shaft or bolt? Some very heavily loaded situations have only taken a few load applications while others have slowly progressed for years and years. However, some insights about them are:

- Once a fatigue crack gets big enough to be easily seen by the naked eye (more than 0.5 mm [0.020″] or so) it always grows perpendicular to the plane of maximum stress. This characteristic is invaluable for its ability to point to the source of the load that caused the failure.
- Most fatigue failures we've seen on components such as pump and motor shafts have taken less than 20,000,000 total cycles from start-up to final failure but there have been some that have taken much longer.
- Under light and variable loads, especially when corrosion is present, it is not unusual to see cracks take 40 to 50 million cycles to propagate across a part. (Corrosion plays this significant role because it continuously reduces

the fatigue strength of the material and, if a cyclical load is present, a crack will eventually start even under very light loads. This results in a situation where there is a long incubation before the crack starts and then an extremely slow-growing fracture, where the crack propagation per stress cycle is almost infinitesimal.)

- Stress or load variations often complicate the diagnosis of fatigue failures. For example, in a machine that sees varying loads, such as a shaft that may turn at 100 rpm, the thought is that it sees 144,000 fatigue cycles per day. But if it is heavily loaded for only 10% of the time and the rest of the time the is lightly loaded, it actually sees about 14,400 significant stress cycles/day.

Before becoming involved with the analysis of fatigue fractures, it would be a good idea to better understand:

- The concept of *stress concentrations* because of their importance in a very high percentage of fatigue failures.
- The effect that metallurgy has on fatigue crack origins.

STRESS CONCENTRATIONS

Stress concentrations, or stress risers as they are frequently called, are features of the part that multiply the local stress. In effect, because of a change in shape or a change in metallurgy, the effective stress in the immediate area of the stress concentrator is greatly increased. Figure 5.2 shows a shaft and the chart below it lists the range of stress concentration factors that are caused by the changes in shape. These higher stresses occur on the shaft surface at the very top of point B and, in an engineering analysis of the shaft, the field stress at point A would be multiplied by the stress concentration factor to obtain the actual stress at point B.

Commenting on Figure 5.2 and Table 5.1:

1. The symbol normally used for the stress concentration factor is K_t.
2. There are substantial differences in stress concentration factors for tension, torsion, and bending loads. This is because the stressed volume of the shaft changes depending on the way the stress is applied. For example, in tension the entire shaft cross-section at B is stressed equally. However, if torsion were to be applied to the piece, only the outermost fibers would see the peak stresses.

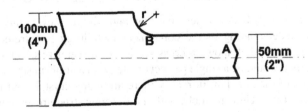

FIGURE 5.2 The cross-section view of a shaft showing a geometrical stress concentration.

TABLE 5.1

Stress Concentrations for the Shaft Shown in Figure 5.2

Radius mm/in	Tension	Torsion	Bending
0.4 mm/0.015"	8	2.5	5.0
1.5 mm/0.60"	3.7	2.0	2.7
3.0 mm/0.12"	2.3	1.6	2.2
6.2 mm/0.25"	1.8	1.4	1.7

Data from Pikey, Walter Pilkey, Ed., *Peterson's Stress Concentration Factors*, John Wiley & Sons, New York, 1997.

3. Stress concentration factors can easily be calculated by many FEA programs. The most complete written text on them was probably done by R.E. Peterson, a Westinghouse engineer, who spent a lifetime compiling invaluable data on stress concentration factors for various component geometries. (I find the book to be a great asset because of the ease of seeing the effect of various dimensional combinations.)

4. An interesting point to realize is that the effective stress concentration factor at the tip of a crack will vary depending on the crack length and the material's properties. The notch sensitivity of the material has to be considered to determine the effective stress concentration factor and there are data that show the effective factor for the tip of a crack may vary from 7 in a very ductile material to as high as 20 in high strength materials.

Some of the most common locations for stress concentrations include:

- Steps or grooves in shafts
- Welds
- Holes
- Keyways and key seats
- On a bolt body, the transition to the threaded section
- Shrink fitted components with sharp corners
- Rough surfaces perpendicular to the stress field

STRUCTURE CHANGES CAUSED BY HIGH CYCLE FATIGUE

Continuing our quest to better understand how and why high cycle fatigue failures occur, a basic understanding of the component metallurgy is extremely helpful. Figure 5.3 is a sketch that shows a cross-section of a metal piece and its atomic structure. Ideally it would be a single perfect crystal with a uniform structure extending to the edges of the part but, in reality there are many crystals and inside each are a number of grains. Unfortunately, within the atomic structure are a tremendous number of *dislocations* and these irregularities are weak points in the structure.

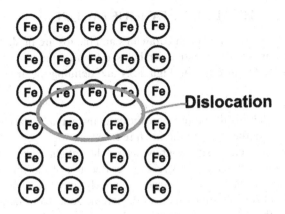

FIGURE 5.3   This shows an idealized iron crystal with a dislocation in the atomic structure.

As a metal is repeatedly stressed the dislocations gradually work their way through the structure and begin to align themselves at grain boundaries. The more heavily the material is stressed, the more rapidly the dislocations move until they eventually line up to create a weakness. With continued cyclical stressing this weak point becomes a tiny crack, which eventually propagates across the structure.

As the crack grows across the part, it starts very slowly but becomes progressively faster and faster until the piece breaks in two. Figure 5.4 is a sketch of a basic fatigue failure face showing a crack that started at the origin and grew slowly across the fatigue zone until it reached the instantaneous (or fast fracture) zone. At that point the crack growth rate instantly changed from a tiny portion of a millimeter or inch per stress cycle to thousands of meters (or feet) per second and the piece fractures.

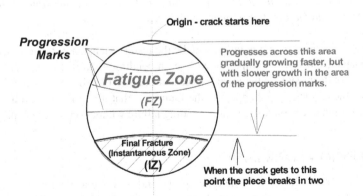

FIGURE 5.4   Showing a basic fatigue failure with the major surface features. This crack growth straight across the part could be the result of a plain bending stress or it could be from repeated tension stresses.

DIAGNOSING A HIGH CYCLE FATIGUE FAILURE

In the subsection below. many of the drawings and photos are of shaft failures, but the diagnostic techniques we mention can be used for all failures.

In analyzing the fracture face in Figure 5.4 and defining the areas and features, one can see:

- *The origin* – This is the point where the cracking actually started and is the oldest and smoothest part of the fracture face.
- From there, the crack slowly grew across the *fatigue zone,* advancing a miniscule distance with each stress cycle. Because of the very slow crack growth, this area of the fracture face is relatively smooth and one clue to the age of the crack is the smoothness of the fatigue zone. Consideration has to be given to the grain size of the material, but the smoother the surface, the slower the growth and the older the crack.
- If the range of the cyclical load isn't constant while the crack is growing, the growth rate and the surface appearance will change, and the result of these load changes are the *progression marks.* These are a calendar detailing the changes in loading over the crack's life. (Some people call them "beach marks" or "stop marks" but the feature describes the growth of the crack across the piece and our opinion is that "progression marks" is a more accurate and descriptive term.)
- When the load on the piece becomes greater than the remaining strength, the piece suddenly fractures across the *instantaneous* (or fast fracture) *zone (IZ).* In this final fracture the failure may be ductile but, in most instances (probably more than 95% of the time in industrial equipment), the instantaneous zone looks like a brittle fracture and the surface is rough and crystalline in appearance. A valuable point is that the size of the instantaneous zone is an indication of the stress on the shaft at the time of the final fracture. *But be careful, because the presence of progression makes tells that the cyclical load has changed over time and the load at the time of final fracture may not be the same as what started the failure.*

The stress that causes a fatigue failure can be the result of tension, torsion or bending. In this chapter, most of the emphasis will be on bending failures. In Chapter 6 there will be a lot of discussion on torsional fatigue and Chapter 12 will cover tension failures in detail. One challenge in diagnosing the failures is that it is difficult to tell the difference between a tension failure and a bending failure without seeing the application.

PROGRESSION MARK BASICS

Progression marks tell us how the crack grew, and they are an incredibly valuable tool to understanding what happened during the life of the failure. The next few paragraphs will introduce the basics, but we will revisit the subject many times.

To understand how a progression mark is developed, visualize a crack growing across a part. With each stress cycle it grows a tiny bit but, if the stress level decreases, the crack growth essentially stops, even though the part is still being cyclically loaded. Later, when the load increases again, the crack resumes growing. This period of no growth results in the change in surface appearance that is called a "progression mark".

(As an aid in understanding progression marks and how they are formed, think about a heavily loaded part with a crack in it. The crack face is growing both rapidly and somewhat irregularly and the crack tip profile looks a little like the coast of Maine with many peninsulas and bays. Then, although still cyclically stressed, the stress level decreases and the growth slowly consolidates into single smooth crack front, like one of Florida's Gulf Coast beaches in the United States. On the cracked part, during the time of consolidation a progression mark is created.)

Fatigue failures are caused by cyclical stresses but if there is no change in the range of those stresses there won't be any progression marks. For example, if the ammeter on a pump motor indicates the load is steady with no changes, there would be no drastic changes in the crack growth rate and no progression marks on the fracture face. But if the ammeter (and the stress level) were varying widely, the crack would grow fast, then slow, then fast, then slow, and there would be series of progression marks that would show an account of the number of variations.

(There are times when these marks can be traced back to specific events in a machine's operating history, giving a better understanding of how and when the failure developed. In one example we looked at a 250 mm [10 in] compressor shaft that had three very definite progression marks over about 5 mm [≈1/4 in] while the rest of the failure face was extremely smooth. The owner was able to trace these progression marks back to one of the roots of the failure being a process upset that put the unit into a surge condition.

(Furthermore, when analyzing the fracture forces and crack growth, knowledge of fracture mechanics is definitely helpful. For example, in high cycle fatigue applications progression marks result from variations in the plastic zone at the crack tip during changes in field stress and, if the crack toughness of the material is known, the exact stress at the time of final fracture can be calculated from the size of the IZ.)

In addition to the process mentioned earlier, there is actually a second mechanism that generates *progression marks*. Depending on the stress on the shaft and the material properties, in the latter stages of a fracture's life, possibly the last hundred or so stress cycles, there are times when the crack grows a visible distance with each stress cycle, resulting in a series of progression marks immediately before the final failure as shown in Photo 5.1. This is a 100 mm (5″) shaft from a trolley drive in a steel mill. A careful analysis showed the cracking started in a fretted area adjacent to the key. It slowly grew across the fracture face then, approximately at the arrows, the progression marks began. These progression marks are also *fatigue striations* as mentioned below and show the crack growth in the last few cycles. (The small instantaneous zone shows that the shaft was very lightly loaded at the time of final failure.)

Occasionally, there is confusion between *progression marks* and *fatigue striations*. *Fatigue striations* show each stress cycle experienced by the part and are usually visible only under extremely high magnification while *progression marks* are visible to the naked eye. Fatigue striations, as shown in Figure 5.5, lie between the progression marks. In some alloys such as aluminum, they are relatively easy to see while in others, such as most of the stainless steel alloys, they are almost impossible to find even

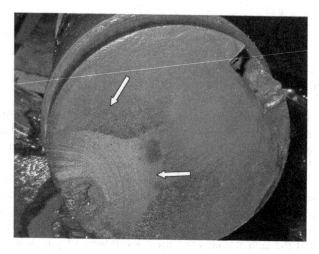

PHOTO 5.1   The fatigue failure of an industrial elevator shaft that started at the keyway, grew slowly across the face, and had a very small final fracture zone. Up until the two arrows there were no visible progression marks.

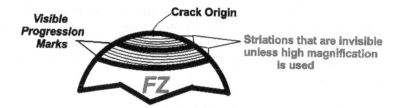

FIGURE 5.5   Fatigue striations show each stress cycle and lie between the visible progression marks. Except for the last few cycles of the failure, high magnification is needed to see the striations.

with an electron microscope. (We should repeat that the progression marks referred to in the previous paragraph, the ones that occur in the last hundred or so cycles "immediately before the final failure" should also be considered as fatigue striations.)

If there is a real need to understand how long it took between fracture initiation and final failure, a common technique is to look at the fracture face using an electron microscope, usually between 500 and 3000 X, and scan the fracture surface. When fatigue striations are found in several areas the crack growth rate across those areas is calculated and integrated across the entire fracture surface, allowing an approximation of the growth rate and time from initiation to final failure. (However, in reality, the process is difficult and, as mentioned above, the striations are sometimes impossible to find.)

FRACTURE GROWTH AND UNDERSTANDING THE SOURCE OF THE STRESS – ROTATING BENDING VS. PLAIN BENDING

Fortunately, from a diagnostic viewpoint, there is a substantial difference in appearance between a plain bending failure, a tension failure, and a rotating bending failure.

In a plain bending failure, such as would occur with one way bending of a leaf spring or a diving board, the crack grows straight across the part as shown in Figure 5.4. (The tension failure equivalent might be a loose bolt.) However, if there is a rotating bending load, the crack growth would be asymmetrical and the fracture face would look like that shown in Figure 5.6.

An example of fatigue cycling caused by a rotating bending load could be a motor shaft with a belt load on it, a reducer with a chain drive, or an axle on a truck. Visualizing the stress in the shaft between the load and the first bearing, we see that, as the shaft rotates the stress at every point on the shaft surface varies, first being in compression, then half a rotation later in tension. This rotating cyclical tension, then compression, then tension, then compression results in the dislocation migration that contributes to the fatigue crack initiation.

A comparison between Figures 5.4 and 5.6 shows that:

1. In the plain bending failure, Figure 5.4, the bisector of the IZ points to the crack origin. But in the rotating bending failure, Figure 5.6, because of the change in stress distribution as the shaft rotates, the crack grows unequally and the bisector does not point to the origin. (This angle is shown as 15° but will vary depending on the relative rotating and plain bending loads. For example, if there is a large plain bending component of the fatigue load and a small rotating bending component the angle will be close to zero.)
2. As a result of the unequal crack growth, it is possible to tell the direction of component rotation.
3. The progression marks, if there are any, will mirror the crack growth and will grow unequally.

There are an almost infinite number of variations in the appearance of fatigue failures. But we'll start with a couple of basic examples and gradually introduce more confusion into the diagnostic procedures.

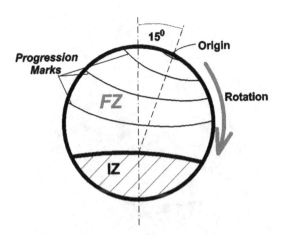

FIGURE 5.6   A rotating bending fatigue failure. Note that the IZ bisector does not point to the origin and the direction of rotation.

Below are photos of two fracture faces. Photo 5.2 shows a portion of the fatigue face that is primarily a plain bending failure. The origin is fairly obvious as are the progression marks that appear to grow straight across the shaft. Also, the change in surface roughness is one of the important tools for identifying that it is a fatigue failure, i.e., as the crack grew across the shaft the surface became progressively rougher because the rate of crack growth was increasing.

(This was a rotating agitator shaft and we would expect it to be a rotating bending failure, but the primary fatigue stress was from a bending stress created by the impeller blades. There was a small rotating bending stress component and a careful inspection shows the progression marks are growing very slightly more on one side than on the other.)

Photo 5.3 is a view of a classical rotating bending failure. Inspection of it shows that:

1. By following the progression marks backward, one can see that the cracking started in the upper left corner of the keyway.
2. The progression marks grew unequally across the shaft and the bisector of the IZ does not point at the origin.

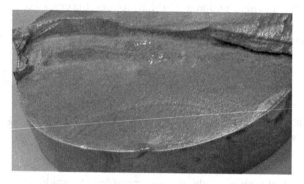

PHOTO 5.2   A fatigue failure with an obvious origin and well-defined progression marks.

PHOTO 5.3   A fatigue failure with clearly defined origin at the corner of the keyway and asymmetrical progression marks that grow unequally across the fracture face.

3. The shaft was rotating in a clockwise direction.
4. The many progression marks show that there were a large number of changes in the shaft stress.
5. At the time of final failure, the shaft stress was very light.

We say the shaft was rotating in a clockwise manner and it is difficult to put into words why the crack grows unequally, but it is easy to visualize with a simple experiment. Take a rolled tube or something similar and rotate it in your hands in the same manner the shaft would rotate in operation. Next, mark a "crack" on the surface of the shaft and perpendicular to the long axis. Then visualize the bending stress in the "cracked" shaft as it rotates. Note how, as you rotate the shaft and the simulated crack rotates into view, the "crack" would have little effect on the stress in the half of the shaft you can see. But as the shaft continues to rotate the effect grows substantially. (The same experiment could be conducted with strain gages to show precisely how the rotationally loaded shaft grows unequally.)

Always look at the failure and look for progression marks or other clues to the shaft rotation. Centrifugal pumps will pump and squirrel cage blowers will blow air even if they are rotating in the wrong direction – and the chance of connecting a three phase electric motor with the correct rotation is 50%.

We have seen many failures where the operators or maintenance personnel "thought it was running in the right direction" but never carefully checked. In one case, a plant had gone through years of performance problems with an air handling system. Then we were asked to balance the fan to reduce the vibration and, when our technician said it was running in the wrong direction, they said, "It couldn't be! Nobody's worked on it for five years!"

Continuing support on this need for analyzing the direction of rotation ... A good friend of ours who was a large motor consultant for an electrical equipment manufacturer once said that over 40% of the motor failures that he had seen [and he defined these large motors as those with separate cooling fans] had their squirrel cage fans running in the wrong direction. Wrong direction of fan rotation results in less cooling air. Reduced cooling air results in higher winding temperatures and, following to Arrhenius' Rule, more rapid insulation degradation and shortened motor life.

PROGRESSION MARKS AND VARYING STRESS LEVELS

An important point that deserves repeating is that the progression marks should be always be examined to understand how the fracture forces changed during operation. As was mentioned, in a plain bending failure the progression marks will grow essentially uniformly across the piece while a rotating bending failure shows asymmetrical growth. If the forces on the part change during operation the shape of the progression marks will also change. Figure 5.7 shows a shaft with irregular progression marks, i.e., they first grow asymmetrically, then they grow straight across the shaft. Inspection of the failure face shows the cracking began primarily due to rotating bending then, as the crack progressed half-way across the shaft, the operating characteristics changed and the driving force behind the crack growth was plain bending.

The importance of carefully inspecting and understanding the progression marks may be difficult to understand but the following example will illustrate it.

The fracture in Figure 5.7 was actually one of a series of a half dozen failures from a group of very similar machines and analysis of the six fractures showed almost identical characteristics. The bisectors of the IZ did not point at the origins and in the early stages of the failures the progression marks were growing eccentrically. Yet later on, as the fracture reached the halfway point on the shaft, the progression marks began to grow straight across the shaft face.

The shaft was on a cutting machine and a sketch of the machine is shown in Figure 5.8. It was used to precisely trim a consumer product and had a blade that struck the anvil roll once per revolution. The plant engineers and supervision were positive that incorrect parts on the cutting roll caused the failure. They went over and over the very detailed setup procedures with their personnel and spent hundreds of thousands of dollars repairing machinery and trying to solve the problem. But the setting of the zero backlash gears was left up to the operators!! Inspection of the

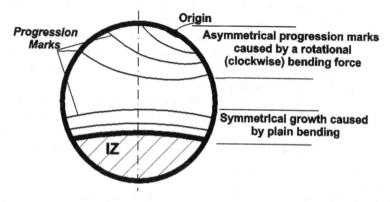

FIGURE 5.7 A fatigue failure face where the progression marks tell that the driving failure forces have changed over time.

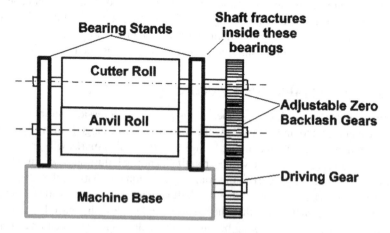

FIGURE 5.8 The plant had several of these machines and had broken six shafts inside the right bearing housing. Then analysis of the fracture face showed the primary cause of the failures was the stress from improperly adjusted zero backlash gears.

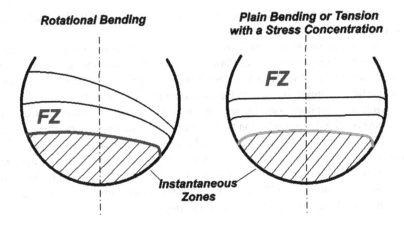

FIGURE 5.9 Rotating bending shafts and plain bending shafts with a stress concentration can have confusing progression marks. However, the rotating bending progression marks will always grow unequally.

progression marks in Figure 5.7 tells us that the forces that started the failure were the rotating bending fatigue forces that could *only* have come from the gears and the plain bending forces from the trimming operation didn't come into significant play until half of the shaft was gone!

PROGRESSION MARKS AND STRESS CONCENTRATIONS

An often-confusing feature of progression marks comes when stress concentrations are present in a plain bending failure. In Figure 5.9 are sections of two shafts with examples of progression marks. Both shafts show progression marks that are concave. The difference is that the rotational bending progression marks are eccentric to the bisector, while the plain bending failure has marks that are essentially symmetrical.

Thinking about a plain bending failure, as the crack grows across the shaft (or other part) the stress is essentially uniform across the fracture face and the crack progresses in a relatively straight line. But with both rotating bending failures and with cracks that grow in a plane with a high stress concentration, the effective stress is higher at the surface of the piece, and the crack grows faster on and close to that surface. This leads to cracks that grow much faster at the edges and progression marks that turn downward at the edges of the high stress concentration.

Two places where this type of progression mark is common are fasteners – with a high stress concentration at the first thread – and shafts that have a sharp groove machined in them.

RATCHET MARKS

If a piece is repeatedly stressed to a level just a little above the fatigue strength it will eventually develop a fatigue crack and, because it was lightly loaded, the cracking will start in only one place. Photo 5.3 is an example of this and the cracking started at the upper left corner of the keyway, i.e., the one location where the fatigue strength was exceeded.

However, if the part were more heavily stressed, cracks will start in several places almost simultaneously as shown in Photo 5.4 where the arrows point to three separate crack origins. (As the effective stress increases there will be more places where the local fatigue strength is exceeded at approximately the same time.)

Looking at this photo one can see that in between these three origins are two *ratchet marks*. The three cracks don't start on precisely the same plane along the axis of the shaft and the *ratchet marks* are the boundaries between fracture planes. As the cracks grow inward and the fracture planes unite, the ratchet marks disappear.

The presence of many ratchet marks is usually an indication of high stress concentrations. For example, in Photo 5.4 the setscrew marks are the stress concentrations that caused the fractures to start at those points.

However, be aware of red herrings! The setscrew marks in the shaft above were only a minor contributor in causing the failure. The black mark to the lower left side of the photo is a scarf mark from burning a bearing off the shaft. When the bearing was burned off, the unequal heating of the shaft caused it to permanently bend and the bend resulted in fatigue cycling of the shaft as it rotated in operation. We won't say it is impossible to cut a bearing off a shaft with a torch and not bend the shaft; it's just that, despite many attempts, I've never seen it done. Cutting bearings off a shaft with a torch is a common technique in many plants and the only way to way to accurately tell if the shaft has been bent is to check the shaft runout before and after cutting.

Most fatigue cracks have some readily visible ratchet marks. In some cases, such as when they appear after the crack has been well established, they don't tell us very much and can be ignored, but there are other times when we can learn a great deal from them. Two examples of this are:

- Frequently, the ratchet mark growth direction is important. When two adjacent ratchet marks grow in different directions the primary origin lies between them. Photo 5.5 shows a gear tooth. The roughness of the surface

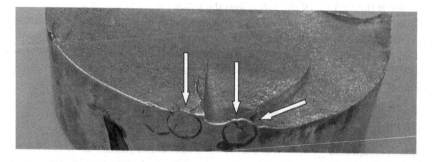

PHOTO 5.4 The cracks started at the stress concentrations from the set screw imprints and then grew inward on three planes that are separated by the ratchet marks.

PHOTO 5.5   A broken gear tooth with ratchet marks growing in opposite directions. From that, we know that the crack origin is in the space between the two ratchet marks, and it's marked with the arrow.

shows that the crack grew rapidly and that it is a low-cycle fatigue failure. Further inspection shows that the ratchet marks in the center of the tooth grew in opposite directions and the primary origin lies between them along the upper edge of the tooth at the arrow. (If a detailed metallurgical analysis were needed, this is where it should be done.)

- When a failure has a great many ratchet marks, suspect high stress concentrations.

An example of a high stress concentration and many ratchet marks can be seen in Photo 5.6 showing a small portion of the fatigue failure face of a 450 mm (18″)

PHOTO 5.6   This shows the OD of a large kiln trunnion shaft that has been built up by welding. The white material toward the top of the photo is some debris and just below that is a layer of weld metal.

diameter kiln trunnion roller shaft. The face has been wire-brushed clean and the many ratchet marks are readily visible. They show that there were numerous failure origins and are indicative of either a very high operating stress or high stress concentrations. However:

A. The rachet marks actually start below the surface, showing that the material varies with a different layer on the surface.
B. There are no progression marks – indicating that the stress on the shaft was constant. (There are some gouges and other surface damage that can be seen in the photo.)
C. The IZ on the shaft was very small, less than 130 mm (5 in) across and less than 7% of the total shaft area. With no progression marks, that shows the load that initiated the failure was relatively small.

Why did the shaft fail? They had had a plain bearing failure that damaged the shaft surface so they had it built up and restored to the original diameter by welding, but they didn't look at the metallurgy. Without the proper preheat, the weld metal left hardened nuggets along the weld/parent metal boundary with residual stresses. Then, with the addition of the normal operating stresses a series of fatigue cracks occurred.

ROTATING BENDING FAILURES WITH MULTIPLE ORIGINS

When a part, like a motor shaft, is subjected to a rotating bending fatigue stress and the stress is more than the fatigue strength, it eventually fails. If the load is relatively light, the fatigue strength is exceeded at only one weak point, cracking starts at only one place, and there is a single origin as shown in Figure 5.6 and Photo 5.3.

However, if that same rotating piece were loaded more heavily the cracking would start at many points around the perimeter and the appearance changes drastically as can be seen in Photo 5.7. This shows a rotating bending fatigue failure where cracking has started around the entire perimeter. The crack grew inward and left a relatively small instantaneous zone. This photo is a good example of several features of fatigue crack analysis in that:

• There are no progression marks, showing that the load didn't vary during the life of the crack.
• The instantaneous zone is small indicating that the load was relatively light.

In view of the fact that the load is light and unvarying, the question is "Why did the piece crack?" and an inspection of the part shows a sharp radius at a step in the shaft causing a high stress concentration. With the actual stress being the load stress multiplied by the stress concentration, the effective stress at the step in the shaft was well above the fatigue strength and cracking began in many places.

STRESS AND STRESS CONCENTRATIONS

Photos 5.8 and 5.9 show two shafts that came off the same 900 rpm motor installation. The first one ran for 24 hours while the second one lasted only 12 hours.

PHOTO 5.7   This is an example of a rotating bending failure with many origins, no progression marks, and a light final load. (A careful observer will also notice that the shaft shows evidence of a weld repair. However, the major problem was a razor-sharp radius at the step in the shaft.)

PHOTOS 5.8 AND 5.9   Two 100 mm (4″) shafts off the same 200 hp motor location, with the left one lasting 24 hours (1.25 million revolutions) and the right one less than 12 hours (600,000 revolutions).

Comparing them we can see the difference between the field stress level on the shaft and the effect of stress concentrations.

Looking at the shaft on the left (and ignoring the smearing):

A. The instantaneous zone is very small, showing there was a light load at the time of failure.

B. There are no progression marks, showing the loading was constant.

C. There are a huge number of tiny ratchet marks, with a fracture origin between each pair.

D. The fact that there were a great many fracture origins with a relatively light load indicates there is a very high stress concentration on the OD.

Next, looking at the shaft on the right, we see:

A. The instantaneous zone is much larger, indicating the load was much greater at the time of failure.
B. There are no progression marks, again showing a constant load.
C. There are fewer ratchet marks and fewer fracture origins.

Comparing the two, we see the shaft in Photo 5.9 has the higher load, but less in the way of a stress concentration. If we had the two shafts in our hands you could see that the radius at the shaft step on the first failure (Photo 5.8) was as close to zero as possible while second failure had a radius of about 1 mm (0.04″). Unfortunately, the load on the second shaft was tremendous. (A major contributor was that the motor shaft was on a belt drive and the sheaves were badly worn. With worn sheaves, the belts have to be tensioned more or they will slip, and the additional tension was enough to cause the shaft failures.)

There is no precise formula, but you can see that, if a shaft is lightly loaded and there are many crack origins, there has to be a high stress concentration in order for the fatigue strength to simultaneously be exceeded in many areas. Conversely, if there are only one or two crack origins the stress concentration factor must be small.

FRACTURE FACE CONTOURS AND STRESS CONCENTRATIONS

Another clue to the severity of the stress concentration is the shape of the fracture face. Keeping in mind that the crack always grows perpendicular to the plane of maximum stress, look at Figure 5.10. Notice that the fracture face is concave and the steepness of the fracture curvature, the cup-like shape and the angles at A and B, indicates the stresses have been intensified at the corner and that there is a substantial stress concentration.

As a comparison, in a component where the stress concentration is much less the amount of concavity would be much less. For example, the shaft section shown in Photo 5.10 has a fatigue crack that has progressed almost all the way through.

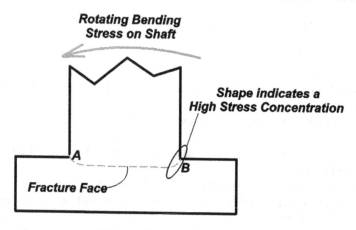

FIGURE 5.10 The cup-like fracture face on this shaft shows there is a very high stress concentration. (Calculation found K_t to be 4.5.)

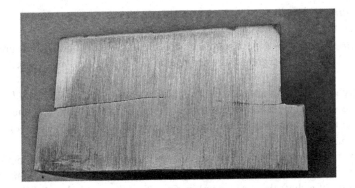

PHOTO 5.10   The relatively flat fracture face on this shaft shows there is a moderate stress concentration and, using Peterson's Stress Concentration Factors, the actual K_t is about 2.1.

Looking at the shape of the crack it is much flatter than that shown in Figure 5.10 because the stress concentration is much less.

INTERPRETING THE INSTANTANEOUS ZONE (IZ) SHAPE

The IZ is the last piece of the part to fail and the shape can show us the forces that were actually acting on the piece immediately before the final fracture. The size of the IZ gives us an idea of the forces at the time of failure and, if there are no progression marks, it is also a good indicator of initiating forces. Furthermore, in Figure 5.11 and 5.12, for the various failure categories note:

- *Plain bending and reversed bending failures* – The shape of the IZ boundary will generally be convex in the last half of the failure. The presence of sharp changes near the outer surface of the piece is an indication of high stress concentrations.

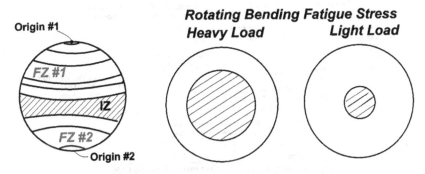

FIGURES 5.11 AND 5.12   On the right is a two-way plain bending failure with the IZ approximately in the center. The larger FZ is usually the older failure, but the progression marks indicate that the fatigue stresses have changed. On the left are two shafts with their IZ's reflecting pure rotating bending with both high and low loads.

- *Rotational bending* – Generally, the higher the total stress, the better centered the IZ.
- *Multiple causes* – A pure rotating load will result in a round IZ. The greater the elongation, the greater the proportion of bending load as a cause.

We've already looked at an example of a plain bending failure with the IZ on one side of the shaft and a rotating bending with the IZ in the middle of the shaft. In addition, if there is two-way bending there is generally an IZ such as the one shown in Figure 5.11A.

However, what happens when there is a combination of rotating and plain bending? The progression mark becomes elongated and Photo 5.11 is an excellent example of the failure of a shaft that has a combination of rotating and plain bending forces.

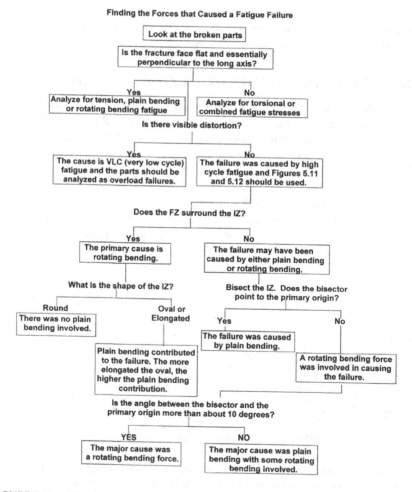

FIGURE 5.13   A basic logic tree to be used in solving the causes of a fatigue failure.

PHOTO 5.11   This shaft tells an amazing amount of information about the causes. We see lots of ratchet marks, lots of progression marks, and an IZ that says the final force was a combination of rotating and plain bending. Also, the progression marks echo that.

GUIDES TO INTERPRETING THE FATIGUE FRACTURE FACE

The first chart, Figure 5.11, is a general guide to the interpretation of fatigue failures. Figure 5.13 is a logic tree that can be used to look at the failures and then Figures 5.14A and B should greatly help in the diagnosis of most fatigue failures.

SOLVING FAILURE

Example 1 – EX5.15

Applying those guides, we find that frequently there will much more distinct ratchet marks than those seen in the three photos above. An inspection of Photo EX5.15 shows some of the features that are seen in attempting to solve field problems. Looking at this stainless steel shaft, we can see:

1. There are definite smooth and rough areas, the FZ and IZ, indicating a fatigue failure. (The change in surface roughness indicates the crack grew at different speeds and proves that the failure didn't happen in a very short time.)
2. The FZ surrounds the IZ, indicating a rotating bending failure.
3. The smoothest parts of the failure FZ are on either side of the keyway, showing that this is where the failure started.
4. The keyway is right up against a step in the shaft, multiplying the stress concentrations and accelerating the shaft failure.
5. The IZ is relatively small showing that the load at the time of final failure was not excessive. (Not only do we see that the final load was low, but we also know from the lack of progression marks that the load did not change over time.)

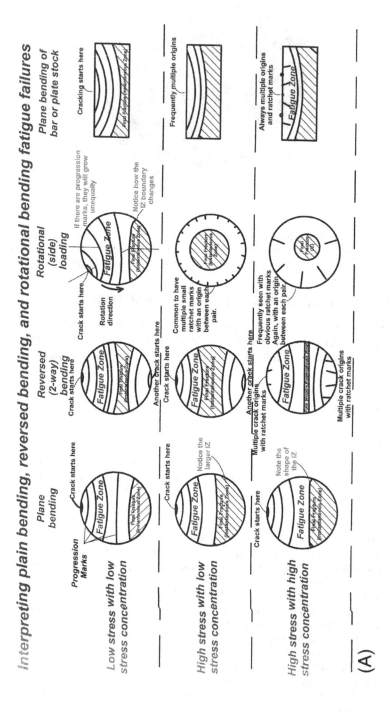

FIGURE 5.14   (A and B) Two diagrams that can be used to solve many fatigue failures.

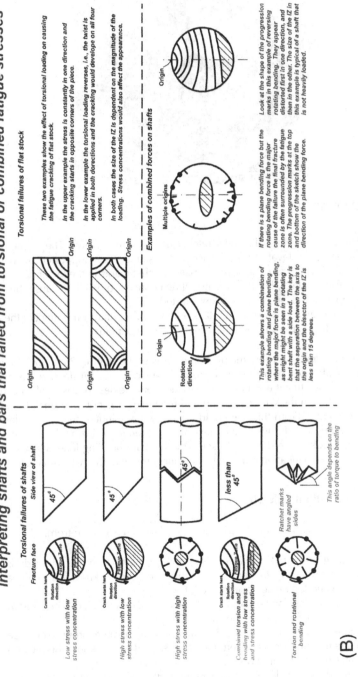

FIGURE 5.14 (Continued)

PHOTO EX5.15

6. There are several substantial ratchet marks at the arrows. They generally indicate elevated stress concentrations, and this is supported by the small IZ. There are also several smaller ratchet marks in the lower half of the fracture face, but they are not of real concern or value because they happened long after the cracking started.
7. The IZ is oval, not round. This tells us that a substantial component of the forces that caused the failure were from plain bending.

In summary, after a very few minutes of analysis of the shaft in Photo EX5.15, we know the shaft failed over time and was not the victim of an instantaneous overload. We also see that it was not heavily loaded, but the design was weak because the keyway was in the wrong position and the radius was not only small but also poorly machined. From this, we could in a very few minutes change the design drawing and prevent future similar failures. However, in analyzing the design, we should also look at the drawing to see why the plain bending load was present, i.e., is there a specification that states the concentricity of the two surfaces?

Example 2 – Photo EX5.16

Photo EX5.16 is actually a bolt, and it is a good example to learn from.

1. Looking at the failure, we can see that the portion on the right side is very smooth while the far left has a rough area, so we know it is a fatigue failure and took place over some time.
2. Then, looking for the smoothest area, we know the cracking started about at the four o'clock position.
3. There are many very fine ratchet marks and a curvature around the edge of the fracture surface where the crack started, so we know there was a high stress concentration.
4. Next, there is a definite discoloration in the area from the origin to the thin dashed line. That's oxidation and tells us that the bolt was cracked

PHOTO EX5.16

for a long time. In that discolored area we also see some progression marks showing the stress was changing over time. But one of the questions we have to ask is "Why did the crack progress about one-third of the way across the part, and then stop growing for a long time?"

5. Going across the photo, we see that the fracture face is essentially flat and perpendicular to the long axis and that there is no distortion or elongation, so we know the failure was caused by tension or plain bending fatigue.
6. The progression marks grow straight across the shaft, so we know there was no rotating bending involved.
7. The final fracture zone, the IZ, wasn't abnormally large or small, so we know the stress levels weren't unusual.

Conclusions and recommendations – This was a bolt failure. With very, very few exceptions, every bolt failure I've seen was the result of inadequate tightening, and this is no exception. However, the corrosion on the fracture face also points to the bolt being in an application where corrosion can't be avoided and some people might claim that corrosion played a major role in the failure. Our recommendation was that the use either a toque wrench of a hydraulic tightener to be sure future bolts had the proper preload.

Example 3 – Photo EX5.17

This was the input shaft for a reducer on a small kiln drive. It was found to have failed at the end of the coupling when they went to start the kiln after a winter maintenance day.

1. The change in roughness, from the very smooth to the very rough, tells that it was a fatigue failure.
2. There is some corrosion on the keyway.

PHOTO EX5.17

3. The main FZ is flat across the shaft and but it's growing eccentrically and that says it was from rotating bending fatigue.
4. The progression marks lead back to the upper right side of the keyway, a common stress concentration.
5. The arrow points to a second fatigue crack that's caused by torsion and it starts at the bottom of the keyway.
6. The IZ is huge and presents a conflict because we have a very smooth fatigue zone which would indicate an old and slow-growing crack, the result of a relatively light load. Yet we have an IZ that says there was a tremendous load.

Conclusions and recommendations – Looking at the shaft, we know there were fatigue forces on the coupling and the fact that the torsional failure starts at the bottom of the keyway leads us to suspect that the coupling was loose on the shaft. (We know there was corrosion present but look at the coupling fit portion of the shaft to see if there is fretting, which will also reduce the fatigue strength.)

The next question is "How can you get rotating bending stress on a coupling shaft?" About the only answer I know of is that the coupling must have been very poorly aligned.

But the big question is "How can you have a fatigue crack that says it's an old and lightly stressed shaft and have a huge IZ?" The answer is that something happened while the machine was shut down, i.e., they accidentally bumped the cracked shaft with a fork truck while doing the maintenance, and that snapped the shaft off!

Example 4 – Photo EX5.18

This was a shaft off a paper machine and the initial diagnosis is pretty straightforward.

1. It is a primarily a plain bending fatigue failure because the crack seems to grow pretty straight cross the face. There is a faint progression mark (at the broken line) about one-fourth of the way across and it is relatively straight, but not quite parallel to the IZ.

PHOTO EX5.18

PHOTO EX5.18A   A is a cross-section view of a similar shaft that has been cut in half and then etched it with a weak acid. Looking at the photo, you can see the layers of a different metallurgy on both sides of the shaft, proof that it has been welded.

2. There is a second FZ on the left side and it is much smaller.
3. We've enclosed the IZ with a box made from a dotted line and it's relatively small, saying the shaft wasn't heavily loaded.
4. There's what looks like a thin ring around the OD of the shaft!

Conclusions and recommendations – The first thing to think of is that this is a rotating shaft that set in a pair of pillow blocks, but it looks like a plain bending failure! How can there be a plain bending failure of a rotating shaft? What happens if a shaft is bent and is subjected to a rotating bending stress? It depends on the cause of the rotating bending stress and, as it rotates, there will be the constant rotating bending stress but at some place there will be an additional peak plain bending stress from the bend in the shaft.

Then we look at that outer layer and see that it is a different metallurgy (see Photo EX. 5.18A). This shaft has been weld repaired, and the heat of the welding has bent the shaft.

They put the shaft in some lathe centers, machined down the fit area, and then welded it, bending the shaft. But because they had located it with the lathe centers, when it was remachined they inadvertently made a shaft with a bend in it. (What they should have done was to support the shaft on the bearing locations when they remachined it.)

Example 5 – Photo EX5.19

This was the drive shaft on a large drum and had been in operation for years. Recently they changed the controls and had had some problems with the drive "hunting". A couple of week later the shaft broke at the edge of the pinion.

1. Again, with the change in surface roughness and the progression marks, it's clearly a fatigue failure.
2. There is some corrosion in the end of the keyway and on the shaft shoulder and that will reduce the fatigue strength of the shaft.
3. The size of the IZ says that the shaft was very heavily loaded at the time of final failure. (This IZ is probably 50% larger than what we usually see.)
4. Of great interest are the progression marks and, looking at them we see:
   a. There were crack origins on both sides of the keyway, but the machine only runs in one direction.
   b. The edges of the IZ show much more rapid growth.
   c. The first progression marks and the early portions of the crack are extremely smooth.
   d. The progression marks are asymmetrical.

Conclusions and recommendations – The very smooth fracture face tells us that shaft has been cracked for a while and the IZ tells that it was heavily loaded, but the really interesting point is that there is asymmetrical cracking on both sides of the keyway, yet the machine always drives in one direction.

Although the major drive force was in the CCW direction, we see that the failure resulted from cracks on both sides of the keyway and was most likely caused by the "hunting" of the drive motor trying to repeatedly accelerate and decelerate the pinion.

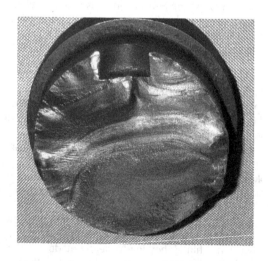

PHOTO EX5.19

Example 6 – Photo EX5.20

This is the shaft that supported a chain sprocket. This is not a new installation and had been in run for quite a while when it fractured.

1. Looking at the face we see the rough and smooth areas that define a fatigue failure.
2. The smoothest portions are really not very smooth and that says the cracking began relatively recently.
3. There are lots of ratchet marks showing that cracking started in many places and they indicate that there was either a high stress or a high stress concentration.
4. The ratchet marks also show cracking started all around the shaft, indicating that there is a rotating bending load.
5. The IZ is rather large and has a greatly elongated shape.

Conclusions and recommendations – The cracking is relatively recent, and we would ask if any maintenance had been performed on the shaft or chain in the last few weeks. (The answer was that they had recently changed the chain and retensioned the drive.)

Really interesting is the shape of the IZ, because:

- If it was just rotating bending, the expected chain load, the IZ should be round.
- But this IZ is the result of a combination of rotating and plain bending loads and the combination is excessive.

Solving what's really happening requires a physical inspection of the chain drive shaft and sprocket, but something is physically wrong. Either this sprocket or the driven sprocket, or the mounting shafts are seriously eccentric.

PHOTO EX5.20

BIBLIOGRAPHY

Atlas of Fatigue Curves, Edited by Howard E. Boyer, American Society for Metals, Metals Park, OH, 1986, ISBN: 0-87170-214-2.

Bannatine, Julie, Comer, Jess, and Hardrock, James, *Fundamentals of Metal Fatigue Analysis*, Prentice Hall, Englewood Cliffs, NJ, 1990, ISBN: 0-13-340191-X.

Metal Fatigue, Theory and Design, Edited by Angel Madayag, John Wiley & Sons, New York, 1969, ISBN: 471-56315-3.

Peterson's Stress Corrosion Factors, Edited by Walter Pilkey, Wiley Interscience, New York, 1997, ISBN: 9780471538493.

Speidel, M.O., "Corrosion Fatigue in Fe-Ni-Cr Alloys", in *Proceedings of the International Conference on Stress Corrosion and Hydrogen Embrittlement of Iron-Based Alloys*. Unieux-Firminy, France: National Association of Corrosion Engineers (NACE), 1977.

6 Fatigue Failures (Part 2): Torsional, Low, and Very Low Cycle, Failure Influences, and Some Fatigue Interpretations

Chapter 5 gives the guidelines for the basic analysis of fatigue failures. Unfortunately, like much of life, there are many complicating differences between the "basic book" and what actually happens. In this section, we'll build on those basics. We will add information on the diagnosis of torsional, low cycle, and very low cycle fatigue failures as well as discuss some of the modifying effects on failure appearances. This chapter will provide you with several more examples and should give you the ability to better solve actual failures.

TORSIONAL FATIGUE AND FAILURES

Torsional fatigue stresses tend to cause some confusion because our human senses can't easily detect them. For example, we can easily see translational vibrations, like a piece of sheet metal vibrating in response to some exciting force or a pump housing bouncing up and down because of cavitation forces, a misaligned coupling, or a weak base. However, we can't easily sense that the shaft in that pump is winding up and relaxing in response to a variable torsional exciting force.

You will find people who have a very hard time understanding the idea of torsional fatigue, but it is present in every rotating machine. Most auto mechanics will realize that one of the reasons there is a flywheel on an engine is to smooth out the power pulses, and anyone who's worked with large steam turbines knows there are rotating speeds that have to be avoided because of torsional resonances. Many people don't recognize those pulsations, but a vibration analysis can be a convincing tool even though it can't measure the actual torsional stress in the shaft. For example:

- On a helical gearset the tooth contact frequency can be measured in the axial direction. That's an indirect measurement of a torsional variation in load.

- On a pump housing the vane pass frequency is another indirect measurement of a torsional force.

When the actual stress values are needed, we have used strain gauges and telemetry or similar systems to detect and understand them.

In order to have a machine shaft rotate, there has to be a torque applied to the shaft, but that doesn't mean there is a torsional fatigue load. Torsional fatigue only comes into play when that torsional stress varies over a relatively short time.

Diagnosis of torsional fatigue failures is similar to diagnosing any other fatigue failure. When the combination of fatigue stress and stress concentrators is low there will be a single origin and the fracture will grow on a 45° angle as seen in Photo 6.1.

An interesting point that can be seen in Photo 6.1 is the smearing of the right-hand piece. When parts break and rub against each other the high spots frequently smear. But if you have both pieces, the high spot on one is the low spot on the other and the pair of faces will give a complete explanation.

This shows a 150 kW (200 hp) ductile iron compressor shaft that failed from pure torsional fatigue. There is a single origin with no stress concentration and the fracture plane is at 45°. Because the shaft is ductile iron and has poor crack toughness, the IZ is both larger and rougher than what would normally be expected from a steel shaft.

If there had been a serious stress concentration on that shaft, the fracture would have developed in several locations almost simultaneously as shown in Photo 6.3. The cracks on this reducer input shaft developed on several of the splines and the failed piece looks almost like a star drill.

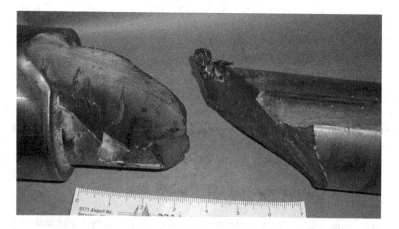

PHOTO 6.1 The torsional failure of a machine shaft showing the classical 45° fracture face. The cracks started at the key seat and the last few progression marks before the final failure can be seen in the left portion.

PHOTO 6.2   Another torsional failure, again with the major fracture plane on a 45° angle.

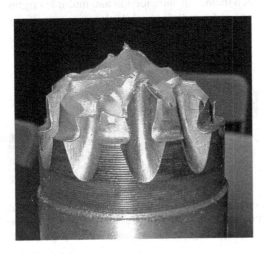

PHOTO 6.3   Another pure torsional failure, but this one has multiple origins and multiple fracture planes resulting in the cone shape. There is some bending but it is small compared to the torsional stress. If there were more bending involved, the angle of the cone would be less, i.e., flatter.

The process of diagnosing the causes of a torsional failure also has to include whether or not there is a bending force involved with that fracture and that involves looking at the plane of the failure as it progresses across the shaft as follows:

1. Single origin
   a. If there is pure torsion involved with a single origin, the crack will grow at a 45° angle.
   b. If there are other forces, bending or tension involved, the angle will be less than 45°, roughly in proportion to the forces.
2. Multiple origins
   a. If there is pure torsion the cracks will grow toward the center and the angle will be even steeper than that seen in Photo 6.3.
   b. If there are other contributory forces, the sides of the ratchet marks will be diagonals, as shown in Photo 6.4, but the fracture face will be flatter.

Photo 6.4 shows a reciprocating compressor shaft where the combined forces caused the failure of an improperly repaired shaft, i.e., there is a layer of weld metal on the OD. The load on the shaft included the torsional driving force, torsional vibration from the irregularities of a reciprocating compressor, plus the weight of a huge motor rotor. (The ratchet marks have diagonal sides indicating that the cracking started from torsional loading, the driving force, in several locations. The main body of the fracture is relatively flat showing that there was also a substantial plain bending load, the weight of the rotor, involved.)

The story about Photo 6.5 is a little hard to believe, but they were from three pumps that sat side-by-side in a pumping station. They all show evidence of corrosion and there are multiple fracture origins on all three, but the instructive point is the difference between them. The pure torsion and multiple origins on the left ended up with that truly conical fracture face, while the center one had a substantial rotating bending component and the fracture plane was much flatter. An interesting point

PHOTO 6.4   This failure started out as a pure torsional failure and the diagonal sides of the ratchet marks show that. But as the failure progressed the loads changed and the bending load became more important in driving the fracture, so much of the fracture face is relatively flat.

PHOTO 6.5   This shows three shafts from one pump house! The one on the left failed from pure torsional fatigue, the one on the right failed from pure bending fatigue and the one in the middle is a combination.

about the one on the right, with just rotating bending, is that the primary origin was 180° away from the keyway.

In the examples above, the failures were caused by loads that were applied along the length of the shaft. An even more common location for torsional fatigue failures starts with the stress concentration at the root of a key seat. In a device that is driven by a key, the driving force is applied across the center of that key and, if there is a typically small radius at the bottom of the key seat, the resultant fatigue stress can be substantial. Both Photos 6.6 and 6.7 show the evidence of torsional fatigue cracking and in both cases loose coupling fits contributed to the failures.

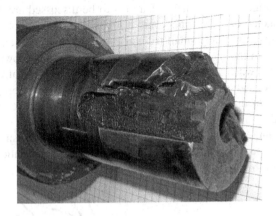

PHOTO 6.6   is the input shaft to a small reducer.

PHOTO 6.7   is the tapered fit coupling end of a large pump. In both cases there was a stress concentration at the bottom of the keyway but, again in both cases, sloppy assembly practices and loose coupling fits contributed substantially to the failures.

One of the problems with loose coupling fits is that the looseness results in fretting and fretting continually reduces the fatigue strength of the shaft material. Eventually the strength drops to the point where a fracture begins. Many mechanics don't like to install couplings with either interference or Loctited™ fits because it entails more work, but the benefit is that the chance for fretting is eliminated and the equipment lasts longer.

RIVER MARKS AND FATIGUE CRACK GROWTH

The last of the fatigue failure surface features to be discussed are of *river marks*. Figure 6.1 is a sketch of a river mark and it looks like the symbol for a river on a physical map. There are many small tributaries to a typical river and, as the water flows downstream in the river, the crack growth also proceeds "downstream". They are actually created when cracks exist on several planes and gradually work together but the value to the analyst is that they show the direction of crack propagation. They occur in all types of materials but are most frequently seen in the fracture of high strength materials or in the later stages of fatigue crack growth.

Photos 6.8 is an excellent example of river marks on a pump shaft failure face. These show the direction of the fracture growth changed dramatically in a short distance and from that we know the forces also changed.

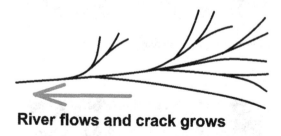

River flows and crack grows

FIGURE 6.1   A sketch of a river mark. These can be used to identify how the crack has progressed across the fracture face.

PHOTO 6.8   Shows river marks on the face a fatigue failure.

PLATE AND RECTANGULAR MEMBER FAILURES

There is a common lament from those folks that don't understand the basics of failure analysis; they want a catalog to show precisely "How to diagnose xxx" with a photo of the exact failure. What they don't understand is that most failure analysis isn't complicated and diagnosing a plate failure, such as that in Photo 6.10, is essentially the same as the diagnosis of any other fatigue cracking.

This is a 50 mm (2″) thick piece of stainless steel is from a large food product plant and looking at it we see:

1. The progression marks and increasing surface roughness show it is a fatigue failure and took place over some time.

PHOTO 6.9   Shows two sets of river marks on the surface of a brittle fracture. This was a large broach that failed due to internal stresses. The river marks show how the cracks grew across the faces.

PHOTO 6.10   A thick stainless steel bar showing a fatigue failure. Across the face are an easily identified FZ, several very clear progression marks, and a host of river marks that show exactly how the cracking progressed.

2. The force that caused the failure is either tension or plain bending and, because the progression marks grow symmetrically for the first half inch (25 mm), it was well centered on the origin.

3. The progression marks show there were a great many load changes during the crack growth.

4. The river marks show the fracture grew from the origin in the center generally toward the lower right showing that as the crack grew larger the force tended to grow a little more off center.

5. The small IZ in the lower right corner shows the part was not heavily loaded.

6. An inspection of the surface at the origin shows the crack started at the base of a fretted pit, reiterating the message in the paragraphs above about the problems fretting causes.

Examining the knife blade in Photo 6.11, we see a fatigue crack with three general origins. The relatively uniform growth and the crack plane again state that the load was either tension or bending. (Usually, determining which of these was responsible is a matter of a simple visual inspection. For example, it would be very difficult to get a tension load on a knife blade.) But two differences between this and the previous failure are:

• The off-center location of the origins indicates that either the stress was not uniform across the blade or there were some stress concentrations toward the tip.

• The large IZ indicates the blade was highly stressed.

This knife failure is really an excellent example of how easy and valuable a field failure analysis can be. The person replacing the knife should be able to *instantly* look at the operation and know that it was either an installation error or a structural problem. There are those who might say "But wait, couldn't it be a metallurgical problem?" The answer is that there might be some metallurgical weaknesses, but it can't fail from fatigue if it isn't stressed excessively, and you can't get tension or bending on a knife blade if it is properly supported.

PHOTO 6.11 An industrial knife blade with fatigue cracking. The knife shaved product off a drum and the fact the it cracked shows they either it wasn't properly assembled to the support or that the support structure was weak.

FATIGUE DATA RELIABILITY AND CORROSION EFFECT ON FATIGUE STRENGTH

In Chapter 3, Figure 3.9 is an S-N curve and it shows the fatigue strength of a metal when stressed at a given rate for a number of cycles. Although there is a tendency to believe that the material will always behave exactly as the curve shows; Figure 6.2 gives a more realistic view of an S-N curve. From this we can see that, even though a piece may be loaded 10% less than the catalog fatigue strength, it may still fail. Meanwhile, another seemingly identical piece which is stressed to 10% more than that fatigue stress doesn't break. The benefit of this chart is that it shows the variability in fatigue data and there are many applications where the range between 1% survival and 99% survival is an order of magnitude. Further complicating this situation is the fact that the fatigue strength can be greatly affected by the operating conditions.

For example, we know that corrosion decreases the fatigue life of a shaft and we know generally:

- The greater the fatigue strength of the material the more rapidly it is reduced.
- The longer the part is attacked by the corrodent:
  - The greater the decrease in fatigue strength.
  - The greater the variability in fatigue strength.
- The more severe the corrodent the more the strength is decreased.

But when we think about an 1800 rpm machine rotating 946,000,000 revolutions per year, we can understand why there isn't a lot of data on the specific effects of various corrodents.

(Our general approach to solving corrosion fatigue problems is to try to eliminate the corrodent. Many times, that is impossible, and we turn to coatings or tapes to try to prevent corrosion from occurring and we have had excellent results from using both anticorrosion tapes and high-quality electrical tape. [We are very careful about changing materials unless we have absolutely positive data as to the full range of operating and shutdown environments.])

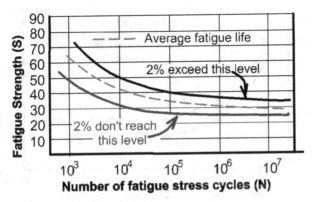

FIGURE 6.2 The graph shows the expected fatigue tolerance bands for a given material. The actual fatigue life for apparently identical components in the same application can vary by as much as a factor of 10.

The magnitude of this decrease in fatigue life is difficult to understand until you look at data such as that shown in Figures 6.3 and 6.4. The first shows the tremendous effect on the fatigue strength of a 1.7 GPa (250 ksi) maraging steel that has been subjected to seawater at 25°C. One would expect the fatigue strength to be at least 600 MPa (90 ksi) in clean dry air, yet the data shows that in seawater, it isn't even 10% of that after less than 10,000,000 cycles – and that's only about a four-day run on an 1800 rpm machine!

Even stainless steels can have seriously reduced fatigue strengths in the presence of corrosion and some of that data can be seen in Figure 6.4.

Most of the data we've seen for mild steels has indicated that after 10^8 cycles in even mildly corrosive water the fatigue strength has been cut at least in half.

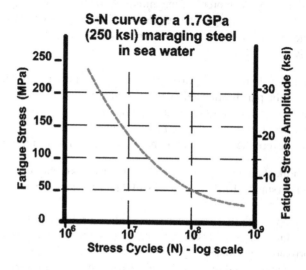

FIGURE 6.3   A chart showing the reduction in fatigue strength of a 1.7 GPa (250 ksi) maraging steel in seawater.

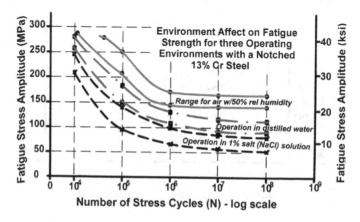

FIGURE 6.4   Showing the effect of atmosphere and corrosion on the fatigue strength of a martensitic stainless steel.

RESIDUAL STRESS CONTRIBUTION TO FATIGUE CRACKING

A shaft or a beam doesn't care about the source of the stress operating on it. The loading could be from mechanical drive stresses, thermal stresses from temperature differentials, or residual stress from fabrication or treating processes. All the part knows is that the actual operating stress should be well below the yield strength for reliable operation.

There are times when residual stresses are beneficial, such as the compressive stress that results from thread rolling on a fastener or shot peening on a crankshaft. The compressive stresses tend to offset the tensile stresses that cause the failures and the compressive stresses greatly increase fastener fatigue strength. Unfortunately, many of the residual stresses we see in plant machinery act to increase the tensile stresses and they frequently contribute to fatigue failures. For most plant machinery, the more common sources of residual stresses involve heating and cooling such as casting, welding, and heat treating.

> The residual stress from improper tempering was the cause of the cluster gear failure mentioned in Chapter 3 and this is a good time to reiterate that, after you solve the physical cause of the failure, you have to look into the human and latent causes. If nothing is changed, the failure will reappear!

One of the cardinal rules of analyzing a failure is that the crack always grows perpendicular to the plane of the maximum stress and an example of how residual stress can contribute to a failure is shown in Photo 6.12. This is a rocker arm out of a racing engine that that has a pair of fatigue cracks in it above and to the left of the bore. During normal operation the cracked area of the rocker arm should never see any tensile stresses, i.e., the push rod pushes upward on the right side and the valve spring pushes upward on the left side. The lower portion of the rocker arm and the area around the mounting stud will see tension but the cracked area should only see compressive forces yet it has two fatigue cracks.

PHOTO 6.12   A fatigue crack in an area that should only see compressive stresses and the result of a heat treating error.

The metallurgical analysis found that area of the rocker was left in tension after heat treating. Then the normal operating stress put it into compression and when the pushrod retracted it went back into tension. It saw tension 6000+ times per minute, then compression, then tension, more than 6000 times per minute, and eventually suffered a pair of fatigue cracks.

We frequently see failures where the residual weld stresses have contributed to originating a fatigue crack, and Chapter 4 has an example where the metallurgical changes due to poor welding procedures caused a brittle fracture.

COMBINED FATIGUE AND STEADY STATE STRESSES

There is a tremendous amount of data on the fatigue strength of a metal and more data on the tensile strength of that same metal, but it isn't easy to understand how the two have be combined in an analysis. For example, think of a pressure vessel that sees changing loads from normal operating stresses, but also has a load from pipe stresses or structural supports.

The first work on this was by Gerber in the late 1800s using some of the data from Wohlber's early fatigue experiments. It generally remained a theoretical curiosity until aerospace applications in the 1940s and 1950s placed a premium on understanding exactly how they affect each other. It became a common aerospace technique in the 1960s but didn't generally reach the rest of the world until computers and Finite Element Analysis (FEA) arrived.

There have been a variety of ways to approach solving the combination of these forces and Figure 6.5 is called a modified Goodman diagram. (Goodman's approach improved on the earlier work of Gerber and Soderberg.)

The horizontal axis, the *mean tensile stress,* is a combination of all the steady state stresses operating on the part, plus one-half of the absolute value of the fatigue stress. In practice the engineer would calculate these stresses, and then draw a

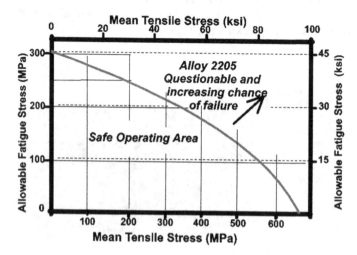

FIGURE 6.5 This is a modified Goodman diagram for the duplex stainless alloy 2205 and shows the safe operating area for the combined fatigue and steady state stresses.

vertical line from that value to the *Goodman line*. Where the vertical line intersects the Goodman line a horizontal line is drawn over to the vertical axis, the *fatigue (alternating) stress*. This is the maximum safe fatigue stress on the assembly and as the mean tensile stress is reduced the amount of fatigue stress can be increased. In Figure 6.5 it can be seen that even if stress relieving only reduces the mean stress by a small amount the probability of failure is decreased.

BASE MATERIAL PROBLEMS

As mentioned earlier, it's not unusual to see folks blame failures on material defects and this certainly happens at times, but it is not very common.

When we started doing failure analyses many years ago, the percentage of base material failures was much higher than it is today and the quality of materials from all sources has improved steadily. As more and more business has involved world-wide competition, international standards have become a much more substantial driving force and the quality of base materials has continually improved.

Less than 1% of the failure analyses we have worked on have involved defects in the original shaft or billet material. Failures in castings and heat-treated pieces are a little more common and weldment failures involving human errors are much more common.

One area where people particularly tend to blame materials has been those that happen at slightly elevated temperatures. It isn't unusual to hear that the cause of the failure of a part that operates at 250° or 300°C (~475° to 575°F) attributed to the elevated temperature and the weakness of the metal. But in reality, the metal strength doesn't drop substantially until well above that (Figure 6.6).

A second common concern for people not schooled in material analysis is the temperature of the parts. Many times people will put their hands on or feel the heat from a piece and decide that elevated temperatures are the cause of the failure when in reality, as shown in Figure 6.5, elevated temperatures don't substantially reduce the fatigue strength of a steel until well above 275°C (500°F). There is more data on

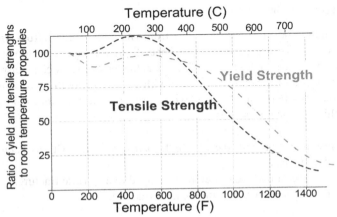

FIGURE 6.6   This shows how temperature affects the yield and tensile strengths of a typical mild steel.

the effect of temperature on material strengths in Chapter 3 and the beginning of Chapter 4, and one thing to be cautious about with hardened components is that they are not operated at temperatures that might alter their heat treatment.

VERY LOW CYCLE AND LOW CYCLE FATIGUE

Most fatigue failures we seen happen from high cycle fatigue and involve millions of cycles between the application of the forces and the catastrophic failure. However, in looking at some fatigue failures it is apparent that they did not take a very long time. For example, if you take a paper clip and bend it back and forth it will eventually break. Sometimes it only takes four or five bends and other times it takes 25 bends. In both situations the paper clips fail from fatigue, but it has taken a lot less than the millions of cycles we discussed in Chapter 5 on high cycle fatigue.

In the beginning of Chapter 5 we defined the regions of VLC, LCF, and HCF. Although the definitions are not very clear, there is a metallurgical difference in the failure mechanisms.

With HCF failures, the fracture develops after the atomic dislocations line up and create a plane of weakness and the crack propagates across the piece leaving very little in the way of deformation. Researchers have found that, typically, about 90% of time for an HCF failure is spent on this crack initiation leaving only 10% for the actual fracture growth.

With VLC fatigue as the part is flexed there is visible deformation and the grains of the part are literally torn apart. The result is a fracture that takes relatively few cycles, almost always less than 25, and the reassembled pieces show some slight deformation.

As with other failures, the diagnosis of a VLC failure involves a close inspection of the fracture face. Causing confusion is the fact that these failures occur at loads close to what would cause an overload failure, i.e., at or slightly above the yield strength and sometimes close to the tensile strength. As a result, the relative ductility and/or brittleness of the material will greatly affect the final appearance.

VLC in Relatively Brittle Materials

Photo 6.13 shows a portion of a broken fork from a medium sized lift truck. Inspecting it one can see that there is a small fatigue zone all the way across the top of the fracture face with the remainder a very large instantaneous zone. A close inspection of this FZ shows at least seven origins, each with a series of four or five semicircular progression marks. Forks are designed with reasonable safety factors and we can tell that from this that the failure was:

1. Not instantaneous with the first high load application. (It has been overloaded in the past.)
2. Grew in at least five cycles before it snapped like a brittle fracture.
3. Very heavily loaded at the time of final fracture.

Comparing this with the fracture face seen in Photo 4.17 we see that fork has five notch-like cracks that had existed for a long time when it was overloaded and broke.

PHOTO 6.13   A lift truck fork that failed after repeated overloads. There are several crack origins and the final fracture looks a lot like a brittle fracture except that there are the multiple origins and evidence of several heavy loadings.

PHOTO 6.14   A forming from a steel mill that was very heavily loaded. The blackened area is oxidation on an older crack surface and the final fracture is the lower third of the roll.

It is probably a trivial point because they both broke after relatively few stress cycles, however the notches in the fork in Photo 4.17 are absolutely uniform and showed no evidence of a growth pattern prior to the fracture while the cracks in the piece in Photo 6.13 saw at least five overload cycles before fracture occurred, and it is highly probable that the machine was regularly overloaded.

A second example of a VLC fracture can be seen in the steel mill forming roll shown in Photo 6.14. The roll had been used for one short run, taken out of service for a week or so, and then put back in run shortly before it broke. Looking at the photo the initial fatigue zone is the blackened area centered at about 11:30 o'clock and an inspection of the fracture found:

1. An extensive network of grinding cracks on the radius.
2. There was some major cracking with three obvious progression marks during the first run.

3. After the unit was put back in run it was very heavily loaded, and the fracture propagated about 230 mm (9″) in one revolution.
4. That was followed by four progression marks about 20 mm (0.8″) apart, shortly after which the roll broke in two.

From the analysis, we could tell the client that, yes, they did load the roll very, very heavily, and that a major reason why it broke was the stress concentration resulting from the grinding cracks.

VLC in Ductile Materials

The fracture shown in Photo 6.15 looks suspiciously like a torsional ductile overload failure. It is from the input shaft of a large machine and close inspection of the face shows a substantial amount of smearing as the pieces twisted apart. The clue that it is VLC lies in looking closely at the OD of the fracture, the area from 11:00 to 1:00 o'clock and the area from about 6:30 to 7:30 o'clock, where several fatigue origins are evident. In order for these to occur the unit had to have been heavily loaded. Our suspicion is that the failure probably took about 8 to 10 fatigue cycles from the first significant load to final fracture but the question that has to be addressed at the plant site is, "Did this tremendous 'fatigue overload' occur immediately before failure or was it a recurring event?" Did the failure occur from a series of eight or ten immediately sequential loads or were the loads applied at widely differing times?

The shaft shown in Photo 6.16 took longer from initiation to final fracture and really belongs in the category of low cycle (LC) and not VLC but this is a good place to introduce it. Examining this paper machine drive shaft, we see the perimeter is

PHOTO 6.15 The very low cycle fatigue failure of the ductile input shaft for a large machine. One substantial clue to the failure cause is the ratchet marks.

PHOTO 6.16 Inspection of the ratchet marks on the low cycle fatigue failure shows that it failed from repeated torsional stress application. But as the failure progressed, because of the relatively ductile material and the many origins, the failure was comparatively flat.

ringed with ratchet marks showing numerous failure origins. Two thoughts to keep in mind during the diagnosis are:

1. Paper machine designs are relatively conservative.
2. When the part failed the operations personnel were starting the machine section. The start-up procedure consisted of "clutching" the drive in and out until the machine got up to speed.

This shaft has a diameter of about 150 mm (6″). The multiple ratchet marks have diagonal sides indicating they were caused by torsional loads and they were surrounded by relatively smooth areas showing that they had existed for some time. Inspection shows the crack then propagated inward for about 10 mm (0.4″) when the loading changed rapidly and the central 100 mm (4″) was the instantaneous zone.

The diagnosis was that the shaft was turned off due to the start-up procedure that aggravated torsional resonances. The failure analysis indicated that this was probably the second such event and that the plant should revise their operating procedures. (They had an identical failure two years later.)

UNUSUAL SITUATIONS

In the chapter on overload failures, there were several examples where ductile materials behaved at least partially in a brittle manner because of the rate at which they were loaded. Similarly, there are times when VLC fatigue failures have unusual appearances.

One of the most interesting and unusual occurs when there are alternating bands of ductile rupture and cleavage on the fracture face. Photo 6.17 shows the fracture of a 100 mm (4″) low speed drive shaft for a curing oven that was having a problem with the oven chain snagging, i.e., periodically seizing in position, and then

PHOTO 6.17   This unusual LC fatigue failure shows an area of brittle fracture followed by a series of fatigue propagation marks, and finally a huge instantaneous zone.

breaking loose. Looking at the fracture, in the upper left corner of the keyway there are a series of fatigue striations over about 6 mm (1/4″) followed by a 20 mm (0.8″) wide brittle fracture area. This was followed by a 10 mm (0.4″) wide area of progression marks and then the entire shaft snapped off. Interpreting these we see:

1. There was a heavy load that caused the initial 5 mm of fatigue cracking. Also, it was aggravated by the stress concentration at the end of the keyway and the corrosion on the exterior of the shaft.
2. The chain snagged and this greatly increased the load on the shaft, but the chain broke loose before there was enough stored energy in the shaft to propagate the crack all the way across.
3. The extremely rapid crack growth stopped after 20 mm (0.8″) and then slowly continued to propagate by fatigue, developing the progression marks.
4. The chain snagged again and this time the shaft broke.

FAILURE EXAMPLES

To understand how a failure occurs, start with the diagnostic chart Figure 2.4 to understand the basic type of failure. If it is diagnosed as a fatigue failure, then look at Figures 5.11, 5.12, 5.13, and 5.14A and B to understand the forces involved and interpret the surface features.

Example 1 – Photo EX6.18

Inspection – Looking at the failed piece and the sketch we see the significant points are:

A. The fractures began on both sides of the key. We can tell this because there are ratchet marks on both sides of the key and because this is the smoothest part of the fracture face. There is some curvature along the edge indicating a high stress concentration at the origin …

 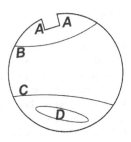

PHOTO EX6.18   This 190 kW (250 hp), 1800 rpm motor shaft is from a paper mill and the figure shows the points referred to in the text. It had been repaired and in run for about six weeks when the shaft broke.

   B. This is a progression mark. It is almost symmetrical, indicating plain bending as the major fracture cause, but notice that there is some distortion off to the left.
   C. Another symmetrical progression mark indicating plain bending. The area between the two progression marks becomes rougher as the crack progresses from B to C. The fact that it is not very smooth indicates a fast-growing crack.
   D. The IZ is relatively small, indicating a light load, and oblong, indicating a combination of plane and rotating bending.
   E. There is some corrosion on the shaft, and that's not surprising in the humid atmosphere of a paper mill.

Conclusions and recommendations – With a motor shaft used for a belt drive we would expect a rotating bending load. However, after the cracking started both of the progression marks are symmetrical and this indicates that the major stress was plain bending. As the failure progressed the shaft became weaker and the rotating bending component of the stress increased. At the final fracture the IZ bisector did not point at the origin, indicating that there was a substantial rotating bending component. But the elongated shape of the IZ indicates most of the loading was from plain bending. Therefore, our opinion is that the major physical cause was that the shaft was bent, and a secondary cause was the rotational loading. (It could have been that the sheave was off center, but that was ruled out.)
   The recommendations to the motor shop were:

   • Improve (or start) a procedure for checking shaft runout before the motor is shipped.
   • Increase the radius at the step in the shaft, reducing the stress concentration.
   • Change the keyway location so it does not end at the step in the shaft. (Reduce the combined stress concentration.)

For the plant we suggested:

   • Checking and recording the V-belt sheave speeds and using them to calculate the tension ratios to ensure the belts were not too tight
   • Being sure to keep to sheaves as close to the bearings as possible, to minimize the overhung loads.

Example 2 – Photo EX6.19

This is a paper mill refiner shaft and the major load should be torsional with very little rotational bending. It operates at 885 rpm and is approximately five years old. Major machine maintenance occurred about two months ago.

Inspection – These fatigue cracks began at two places A and B. However, start by looking at the step in the shaft and we see there is obvious corrosion pitting, reducing the fatigue strength of the steel. Then we see a ratchet mark at A indicating that there were actually two crack origins, one on each side of the ratchet mark. Looking at the progression marks surrounding these origins, they are not symmetrical, indicating some rotating bending is involved. Next, at B there are also two origins centered on a ratchet mark one-third of the way from b1 to b2. The ratchet marks at these two positions are pointing away from each other proving that there is a fracture origin between them. Furthermore, there are several faint asymmetrical progression marks between these origins and the IZ at D. The rough area at C is uncommon but the river marks indicate that it was one of the last areas to fracture before the shaft came apart at D. D is the final fracture and is relatively rectangular, indicating the final fracture was from plain bending.

Conclusions and recommendations – This looks complicated, but it is actually a relatively straightforward failure. We can see that corrosion greatly reduced the fatigue strength of the shaft. Then the cracking started, first at A and then at B. Both of the fracture planes propagated primarily by rotating bending but with a significant component of plain bending until the final fracture when the shaft was not heavily loaded.

Our recommendations to the people at the plant were:

- Improve the shaft alignment to eliminate the rotating bending load.
- Tape or coat the shaft to try to reduce the fatigue strength deterioration from corrosion.
- Add a concentricity specification to the shaft drawing to reduce the plain bending load.

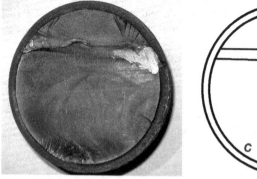 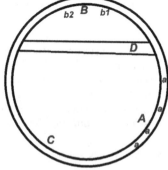

PHOTO EX6.19 The fatigue failure of this paper mill refiner shaft was substantially hastened by the corrosion around the step in the shaft. The diagram to the right shows them major points referred to in the text.

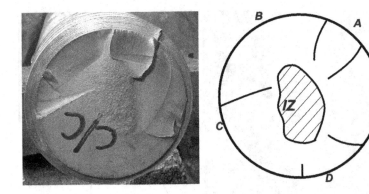

PHOTO EX6.20 This is the shaft on a large gate valve along with a diagram of the fracture face. The shaft doesn't rotate but saw a bending fatigue failure.

Example 3 – Photo EX6.20

This was the stem for a 2 m wide (6 ft) gate valve in a sewage treatment plant and it failed three months after installation. The stem was driven by an electric motor through a reducer and rotated to lift and lower the gate. The gate was positioned vertically and designed to move up and down in a gap between the guides.

Inspection – The examination began with noting that the surface of the machine's stainless steel shaft was very rough. Then, examining the fracture face, there are five general areas of origins at A through E. Along with those origins are some prominent ratchet marks, showing the fractures began on different planes. Next, looking at the progression marks, it can be seen that they progress inward symmetrically from each origin. The final fracture zone is about what would be expected in a piece this size.

Examination showed the stem was being flexed back and forth while in position. The vibration from the flexing allowed the reducer to slowly allow the gate to drop, repositioning the stem and creating another crack.

Conclusions and recommendations –

- The valve stem was rigidly held while the huge gate was loose in the gate guides. The gate guides were not properly installed and allowed the gate to move back and forth, flexing and heavily stressing the stem.
- The rough machined surface resulted in small stress concentrations that hastened the failure.

Example 4 – Photos EX6.21 and EX6.22

This shaft is about 120 mm diam (4 3/4") and rotated at almost 600 rpm. It was used in a process plant and is interesting because of the shape. The sketch shows a cross-section of the piece and the operating details are not as important as under-standing the surface features.

Inspection – Shows cracking began around the entire shaft at the shoulder radius. The multiple ratchet marks point toward a high stress concentration and

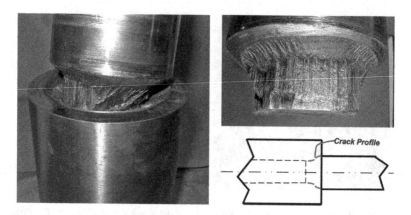

PHOTOS EX6.21 AND EX6.22   These show two views of the failure of a long shaft that had both a large step and a hollow bore. The sketch to the right shows a cross-section with both the crack profile and the hollow bore.

the steeply angled fracture face is perpendicular to the stress field. The early fatigue zone is smooth around the entire shaft. It then grows into a much rougher and faster fracture with numerous river marks, then into an instantaneous zone that is parallel to the shaft axis.

From the fatigue zone it appears that the only stress on the shaft was rotating bending and the multiple origins around the shaft show the actual stress, i.e., the stress × stress concentration, is very high. There are numerous progression marks, and this indicates the load is changing with the size of the IZ indicating it was very heavily loaded at the final fracture.

Conclusions and recommendations –

- Improve the alignment to reduce the rotating bending stress on the shaft.
- Shorten the inner bore, allowing much more room between the stress concentrations from the internal and external steps in the shaft.
- On both steps, make the transition radii larger to reduce the stress concentrations.

The rough, almost woody appearance of the instantaneous zone lead many people to suspect that the material was defective. However, this is the typical appearance of an IZ or a brittle fracture that propagates in the same direction that the piece was rolled. We are used to seeing instantaneous zones that run perpendicular to the rolling direction and they produce rough crystalline appearing surfaces. When they grow in the rolling direction, they tend to fracture along the grain flow lines and the result is something that looks like a piece of petrified wood.

Example 5 – Photos EX6.23 and EX6.24

Both Photos EX6.23 and EX6.24 (and Photo 6.1 earlier in this chapter) show views of an 86 mm (≈3.5″) drive shaft from a process winder. The plant had had problems with short coupling life and had recently replaced the coupling with a larger one. The coupling fit is at the far right side, with the winder toward the left.

PHOTO EX6.23 AND EX6.24   These are two photos of a drive shaft with the angle of the fracture showing that the cause was a torsional fatigue stress.

Inspection – The diagonal fracture face shows this is a torsional failure. Then, examining the surface we see that:

1. The surface gets progressively rougher as it travels across the shaft but even at the beginning it is not very smooth.
2. As shown in Photo EX6.24 the cracking starts at the upper corner of the end milled keyway.
3. There are numerous progression marks indicating many load changes.
4. The final fracture zone, the IZ, is relatively large.
5. There is obvious fretting at the end of the key fit.
6. The coupling fit appears to be tight with only very slight fretting on the drive side of the interference fit.
7. Hardness testing showed the tensile strength of the shaft was about 80,000 psi.

Conclusions and recommendations – This is a pure torsional fatigue failure with a very high torsional load. The coupling fit was excellent and did not contribute to the failure. The fact that the previous couplings had also been destroyed leads us to suspect there is a torsional resonance in the shaft.

We recommend further analysis to understand the shaft loads, i.e., checking to see if there is a torsional resonance and also taking high speed recording ammeter readings on the motor to see the magnitude of the motor loads. With that combination they should be able to see how highly the shaft is loaded and how the loads vary. Possible corrective actions could include changing the torsional stiffness of the assembly by changing the type of couplings and strengthening the shaft but an understanding of the true forces on the shaft is a necessity. Along those lines, we suggested that they look at the variable speed drive controls to ensure that they were not introducing speed variations that could contribute to the varying torsional stresses.

Example 6 – Photos EX6.25 and EX6.26

We were doing a class for a company that had work crews around much of the world when one of the attendees brought this into the room for the hands-on portion of the class. The trailer hitch was stamped with a maximum load rating of 6000 lbs and a maximum tongue weight of 600 lbs, yet their product seldom weighed that little.

PHOTOS EX6.25 AND EX6.26   Two great views of a torsional failure of a flat section.

It is an interesting example of a torsional fatigue failure of a flat bar. We can see origins at each of the four corners, with four relatively small fatigue zones and two large instantaneous zones. It also appears that the fit of the stem of the hitch ball is not snug in the hitch, but that may be the result of deformation that occurred during the failure.

A close examination in Photo EX6.26 shows the fatigue zones are corroded, indicating that the failure wasn't new, and that there is some plain bending fatigue starting at the left interior corner on the lock washer.

Conclusions and recommendations – We didn't have any history on how the hitch had been used however the company's product was used in the mining industry, frequently in field operations. Looking at the fracture, it is obvious that the hitch ball was frequently subjected to significant fatigue loads and that this wasn't a new failure.

A failure at the wrong time could result in a six-figure loss as well as endangering the workforce. We recommended they institute a practice of routine hitch inspections before going out on a delivery and proper sizing of the hitch.

Example 7 – Photos EX6.27 and EX6.28

This is the end of the coupling fit on a shaft that drove the ID fan for a coal-fired in a power plant in Texas.

It had been in run for years at 880 rpm but, in a cost saving project, the plant put a variable speed drive (VFD) on the fan. That that allowed them to run the fan much slower, but several weeks after startup the fan shaft failed.

There was dispute between the people who installed the VFD and the operating group as to the cause of the failure.

Photo EX6.28 is a better view of the fracture face and from it we can see:

- The angle of the fracture tells that it is the result of a pure torsional stress.
- There is corrosion on the face but that is the result of a long and unprotected wait before the piece was sent to us.
- There is fretting corrosion in the keyway.
- The fatigue zone is relatively large but has numerous river marks across it.
- The final fracture zone is relatively small.
- We don't show it, but the elastomer elements of the flexible coupling were also destroyed.

PHOTO EX6.27   shows a slice from a shaft that was sent to us. As you can see, the piece was cut off on an angle and the fracture face is at 45° to the centerline of the shaft.

PHOTO EX6.28   is a view of the actual fracture face and the change in surface roughness shows it's the result of torsional fatigue.

Conclusions and recommendations – From the 45° angle of the fracture face, it's relatively obvious that this is a fatigue failure. Even though the IZ is small, the fracture face with the river marks shows that the failure was relatively rapid.

There may have been some reduction in the fatigue strength from the corrosion on the OD of the shaft, but the major problem was the torsional vibration caused by running the shaft at a torsional critical speed. Our recommendation was that they run the fan at its original rpm until they could shut the system down and install strain gages on the shaft. Then run the fan at varying speeds to determine the critical frequencies.

Example 7 – Photo EX6.29

The photo shows a broken crane hook. At the top of the photo is the latch and from that we know that the tip of the hook lies in line with the latch.

PHOTO EX6.29   Look carefully at the left side of this fracture face, and you can see it has been side loaded more than once!

At first glance it looks like a brittle fracture and people might feel this was from a single overload. But looking closely at the left-hand side of the fracture shows that there are several progression marks.

Conclusions and recommendations – The fracture clearly shows the hook has been repeatedly heavily side loaded. (Remember that the hook was designed with a safety factor of at least three, based on the tensile strength, and that the hook material is ductile.) This is proof of a repeated safety violation.

BIBLIOGRAPHY

Metal Fatigue, Theory and Design, Edited by Angel Madayag, John Wiley & Sons, New York, 1969, ISBN: 471-56315-3.

Osgood, Carl, *Fatigue Design*, John Wiley & Sons, New York, 1970, ISBN: 0-471-65711-5.

Sachs, Neville W., *Failure Analysis Made Simple: Bearings and Gears*, Reliabilityweb.com, Ft. Myers, FL, 2015, ISBN: 978-1-941872-30-7.

Sachs, Neville W., *Failure Analysis Made Simple: Shafts and Fasteners*, Reliabilityweb.com, Ft. Myers, FL, 2018, ISBN: 978-1-941872-81-9.

7 Understanding and Recognizing Corrosion

SOME BASICS ABOUT CORROSION

Corrosion is the deterioration of a metal, almost always due to of oxidation. We've all seen rusty steel parts and generally recognize corrosion but, like a lot of things, why it happens is much more complicated than it appears to the casual observer.

One of the confusing things about corrosion is that over the years we've tended to classify different types by their appearance, but my opinion is that really it helps to think of them as being in two general categories:

- The variations on the basic corrosion cell shown in Figure 7.2, where materials are transformed into oxides. These include galvanic, pitting, crevice, fretting, intergranular corrosion, and selective leaching.
- The mechanisms that cause environmental cracking including hydrogen damage and the forms of stress corrosion cracking.

In this chapter, we'll start with some technical data explaining some of how and why corrosion occurs, and then we'll get into a series of examples that show some common problems and corrections.

Of all the failure causes, corrosion is probably the most significant and is the most expensive. For the past 75 years, studies conducted by the Battelle Institute and other organizations have shown the United States spends somewhere between 3 and 4% of our GDP on corrosion prevention, treatment, and correction. Some recent studies have found it to be about 3.1%, and, if we look at 2017, where the World Bank said the U.S. GDP was 19.4 trillion dollars, the cost of corrosion in the country was more than 600 billion dollars! For most of the developed world the cost is similar however the warmer and more humid countries are even more burdened.

The reason corrosion exists is shown in Figure 7.1. We take minerals out of the ground and expend a tremendous amount of energy to make them into useful metals. This increases their potential energy and Mother Nature spends the rest of the metal's life trying to reclaim that energy. It is interesting to note that the metals found naturally occurring, such as silver and gold, are generally those that have a great deal of corrosion resistance. The rest of the metals we use are mined as oxides.

One of the keys to understanding corrosion is to realize that it is the result of electrical currents and electrochemical reactions. Different types of corrosion cause those currents to be initiated in slightly different manners but a thorough understanding of the current generation and conduction (and the reaction kinetics at the anode and cathode) is a necessity to understand and measure corrosion.

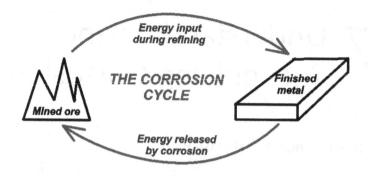

FIGURE 7.1   This shows the basic corrosion reaction. We put a large amount of energy into ores to convert them into useful metals, and then corrosion releases that energy.

Almost all corrosion requires that some moisture be present but there is dry corrosion. It is not very common and almost always happens at elevated temperatures and is called *direct conversion*. A simple example can be seen by heating a mild steel rod with a torch until it is red hot and it immediately develops a flaky gray-black oxide scale. Developing this oxide scale is corrosion.

A more common example of direct conversion can happen with a barbecue grill with a cast iron grate. The food is usually supported on the grate and these grates constantly flake off large pieces of scale, i.e., large pieces of iron oxide that are the result of direct conversion. (My personal solution to this problem was to make a grate from Type 316 stainless steel and the grate is now on its third housing.)

But then think about the hand and toe warmers we use when we're outdoors in cold climates. These are one of the very, very few examples of direct conversion that take place at close to room temperatures and they are technically a direct conversion corrosion mechanism.

With the exception of those direct conversion examples and possibly some of the hydrogen damage mechanisms, one of the more important points to remember when trying to solve corrosion problems is that there must be moisture present in order for the attack to occur.

Figure 7.2 is an example of a typical wet corrosion reaction. It shows a steel bar, but it could be any metal in a conductive electrolyte. For the reaction to take place, there has to be an anode, a cathode, and a liquid is needed to conduct the corrosion currents from one to the other. The anode is attacked and releases metal ions into the liquid to form oxides, while at the cathode the corrosion reaction produces atomic hydrogen, H^+.

In most cases, these hydrogen atoms immediately unite with another hydrogen atom to create the common molecular form, hydrogen gas, H_2. But the fact that hydrogen is generated in the atomic form is the source of many of the hydrogen damage problems.

(In understanding the mechanisms of hydrogen damage, it might be helpful to realize that, on a microscopic basis, steel is to hydrogen about like a wire fence is to feathers. The fence will slow down the feathers, but it won't stop them from passing

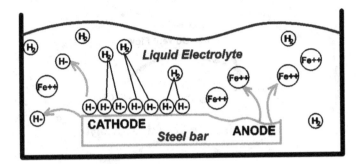

FIGURE 7.2 This shows a view of what actually happens with the corrosion of steel (or any other metal) in a liquid. Note that the liquid is needed to allow the electrochemical circuit to be completed.

through. In a similar manner, the hydrogen atom is so small it can wander through the interstices in the steel. When it finds and joins with another hydrogen atom, the molecular forces result in hydrogen gas which requires a much greater volume, and significant internal forces are created.)

A point to realize about this hydrogen generation is that there will always be some dissociation. That is, although most of the hydrogen atoms unite to form hydrogen gas, some of them will remain as ions. It is these ions that result in the damage. In addition, there are some chemicals, such as sulphur, that tend to restrict the ion's formation of hydrogen gas molecules.

The idea of free hydrogen atoms and that there are always some present can be a difficult one to understand. One interesting example that will illustrate this is the degradation of titanium heat exchanger tubes in a plant where we worked. The vessels were large shell and tube heat exchangers with steam on the shell side and process liquor on the tube side. The tubes were welded titanium and they would start failing about seven years after installation. The failure mechanism was true hydrogen embrittlement, i.e., the free hydrogen atoms from the steam would unite with the titanium and slowly form titanium hydride, a brittle ceramic-like material. Eventually the entire tube wall would be converted to this brittle material and the tube would crack and leak. But the portion of the tube outside the tube sheets, in the water box, was completely unaffected and remained as ductile as a paper clip! The culprits were the free hydrogen atoms in the steam.

Looking at typical corrosion, a conductive liquid has to be present to complete the corrosion circuit. The current travels through the anode and cathode, returning through the liquid. If the liquid isn't present the circuit isn't completed, and the reaction stops. Within that corrosion cell, Figure 7.3 shows how the currents travel.

Some important comments about the liquid are:

1. If the conductive liquid were prevented from getting to the surface there would be no corrosion – because there would be no way to conduct the corrosion current. This is why many corrosion prevention coatings, such as paint, don't let liquid contact the metal.

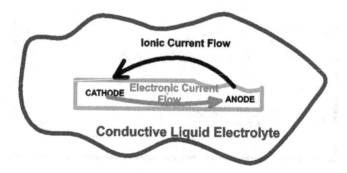

FIGURE 7.3   The current flow in a corrosion reaction.

2. In a similar manner, if the liquid contact with the surface were slowed, the reaction rate would also become slower. With steel a rust buildup slows the contact with fresh corrodent and reduces the corrosion rate. This is the basis behind the chemistry of COR-TEN™ (ASTM A 242 and ASTM A 588) weathering steel where the rust adheres more tightly than on normal steel and prevents the corrodent from getting to the surface.

3. If the liquid contained little or no oxygen, the corrosion rate would be greatly slowed. If there is no or reduced oxygen available, such as in the mud under a lake or a stream, there's greatly reduced oxidation and the driving force for the corrosion reaction effectively stalls.

4. The liquid doesn't have to be water and doesn't have to be at low temperatures. An example of this occurs in coal-fired boilers where pitting corrosion occurs when the sulphur compounds in the coal liquefy and carry the corrosion currents at temperatures well above 550°C (1000°F).

5. Atmospheric corrosion tends to start in nooks and crannies and at surface irregularities. Research has shown that first traces of condensation (and corrosion) generally begin at about 60% relative humidity and get worse at the humidity increases.

CONDITIONS AFFECTING CORROSION RATES

Many conditions affect corrosion rates including temperature, exposure time, moisture and liquid conductivity, pH, oxygen availability, solution contaminants, flow velocity, surface conditions, etc. The reactions are extremely complex and dependent on multiple variables and, because of the number of variables and their interactions, it is often difficult to use laboratory experiments to accurately reproduce the actual field corrosion rates.

TEMPERATURE EFFECTS

Other than the presence of moisture, we can think of nothing that affects corrosion rates as much as temperature.

A good place to start in understanding these effects is with Figure 7.4. This is a chart of the corrosion rate of a metal in water vs. temperature but doesn't show values

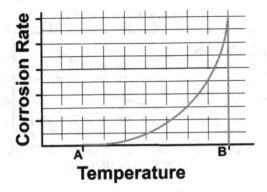

FIGURE 7.4 Showing the effects of temperature on the corrosion rate of a metal at normal atmospheric pressure.

for the temperatures. Looking at it we can see that as the temperature drops to point A the corrosion stops. From this we know that point A is at 0°C (32°F), because the water freezes and ice won't conduct the corrosion currents. Then as the temperature increases and reaches 100°C (212°F) at point B, the water boils and turns to a gas and the corrosion stops again – this time because gases can't conduct the corrosion currents. (An ironic point about this chart is that if we add salt to the water, such as deicing road salts, it adds to the corrosion not only by decreasing the freezing point, but also by increasing the conductivity of the water.)

Looking further into the effects of temperature, it helps to understand the work of Svante Arrhenius, a Swedish Nobel Prize winner who spent a goodly part of his life publicizing the fact that, for most chemical reactions in the range of temperatures around ambient, for a 10°C (18°F) increase in surrounding temperature the rate of a reaction approximately doubles.

(Arrhenius' equation essentially states that the rate constant, a key factor in calculating the rate at which a reaction will proceed is an exponential function of the absolute temperature. Understanding it is helpful when it comes to corrosion rates, but it is invaluable for understanding any of the degradation mechanisms that involve chemical reactions. We refrigerate food to slow the reaction rates. We wash clothes in hot water, not because the detergents can't clean them in cold water, but because the detergents act so much more rapidly in hot water. As temperatures go up, electronic components die more rapidly; polymer products such as seals and belts degrade more rapidly, and corrosion rates almost universally increase, approximately doubling with every 10°C [18°F] increase.)

In Figure 7.5, the effect of temperature on the corrosion of several alloys in sulphuric acid is shown. Again, two valuable points we can glean from this graph are the differences in corrosion rate with temperature and the effects of the different alloying elements. The gray cast iron, even with the relatively high acid flow rate of 5 feet per second does relatively well at 40°C (104°F) but when the temperature increases by 30°C (86°F) the rate of attack has increased by less than what would have been projected. (It starts at 0.06 ipy, goes up 30°C, so that would be a multiplier of 8, for a projected corrosion rate of 0.48 ipy, but the actual rate is only 0.14 ipy.)

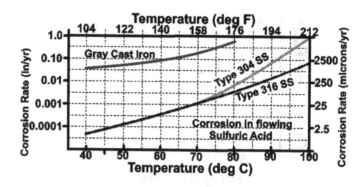

FIGURE 7.5 Showing the corrosion rate of three alloys in flowing 93% H_2SO_4 at various temperatures.

Meanwhile, in the same temperature range, the attack on the Type 304L SS increases at about the same rate, but then at the higher temperatures the corrosion rate increases much faster.

pH EFFECTS

The pH value of a solution is a measure of the relative number of hydrogen ions in the solution and Figure 7.6 shows a pH scale. It is the common measurement used to express whether a material is acidic or alkaline and the scale ranges from 0 to

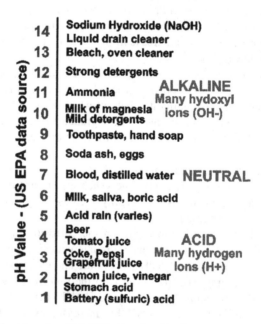

FIGURE 7.6 A chart based on U.S. EPA data showing the approximate pH of various common substances.

14. A low pH indicates the material is very acidic while a high pH indicates that the material is alkaline (basic).

The pH of a corrodent can have a tremendous effect on the corrosion rate of a metal. Normally the lower the pH of a material, the more corrosive it is and acidic materials are usually more corrosive than neutral or basic materials. In general terms, the acid tends to destroy the oxide coating on a metal and increases the amount of corrodent that can reach the surface.

Although corrosion rates drop as the pH approaches 7, in some cases they begin climbing again when the solution becomes highly alkaline. However, neither highly acid nor highly basic solutions are universally corrosive. For example, nickel alloys are highly resistant to very alkaline solutions while the same materials rapidly corrode aluminum alloys. Diagnosing the specific effect of a pH requires research into the detailed conditions.

One example of how both pH and temperature effect corrosion rates is shown in Figure 7.7 and is a good general guide for the combined effect on mild steel.

It's important to understand that this is just a general guide, and both the local conditions and the corrosion mechanism have to be taken into consideration. For example, many industrial sulphuric acid storage tanks are made from carbon steel and, at room temperatures, 93% sulphuric acid steel initially corrodes incredibly rapidly. But it forms a tight layer of ferrous sulfate, which reduces the amount of acid that can get to the steel, and the corrosion rate then drops rapidly to a very low rate. However, if this protective layer is disturbed it exposes the steel and the corrosion continues. In inspecting sulphuric acid storage vessels, the important places to look are:

1. Those areas where there could be flow disturbances such as a fill pipe near the tank wall or an outlet pipe where there is an elevated flow velocity.
2. The areas where the hydrogen bubbles, from expected slow corrosion, could flow together and scrub away the protective sulfate layer. (This damage is frequently seen in the heat affected zones of vertical welds and along the top of horizontal steel pipe runs.)

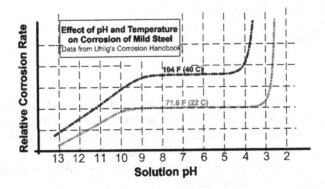

FIGURE 7.7 Showing how pH and temperature can combine to vary the corrosion of mild steel.

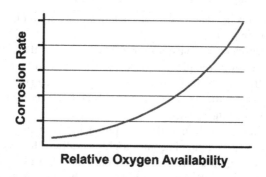

FIGURE 7.8   Oxygen availability is important to the corrosion rate. If oxygen isn't available, an oxide can't be formed.

OXYGEN AVAILABILITY

Figure 7.8 shows a general chart of the effect of oxygen concentration on corrosion rates. Oxygen has to be present for oxidation (corrosion) to occur and the greater the oxygen availability, the greater the corrosion rate.

Some practical examples of the effects of oxygen concentration and availability on corrosion are:

- In many tanks and storage vessels the area of greatest corrosion is in the range of the normal operating liquid level because this is where the wetted surface can readily absorb oxygen from the atmosphere.
- In marine applications the area of greatest corrosion is in the splash zone, with readily available oxygen, while areas deep underwater where there is little oxygen tend to see much less corrosion.
- An unusual example of the importance of restricting the availability of oxygen to both the corrodent and the surface was in a series of large mild steel liquid calcium chloride storage tanks. The tanks used a layer of fuel oil floating on the surface to not only coat the sidewalls of the tank (and restrict oxygen from getting to them as the liquid level varied) but also to restrict oxygen from being absorbed by the solution.
- One of the mechanisms used to reduce the corrosion by boiler water is to deaerate the water and remove the oxygen.

EXPOSURE TIME AND FLOW EFFECTS

In general, corrosion rates tend to decrease with time. Figure 7.9 is a diagram that shows the typical slowly decreasing corrosion rate seen in a laboratory experiment. This was in a solution that wasn't agitated and a buildup of hydrogen gas at the cathode or a buildup of corrosion products at the anode forced the corrosion rate to slowly decrease. (But, in real life, local conditions can affect these build-ups, and the corrosion rate can vary tremendously.)

Another example of how time affects corrosion rates can be seen in the data of Figure 7.10. This shows the results of a long-term analysis of horizontal mild steel

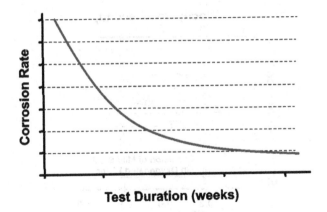

Test Duration (weeks)

FIGURE 7.9 A graph showing how the corrosion rate of steel usually drops off during a laboratory experiment, and Figure 7.10 shows how this also happens in the real world.

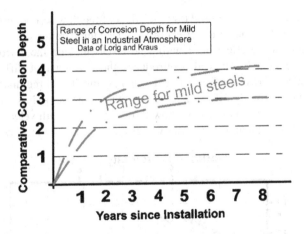

Years since Installation

FIGURE 7.10 Data from a long exposure term showing how the field corrosion rate drops off significantly with time.

corrosion rates in an industrial atmosphere and an important thing to realize is, even though it would be essentially impossible to replicate the experiment's environment, this is what generally happens over a long exposure time. But, if the environment is altered, such as a diligent person trying to clean the oxide off, the corrosion rate will stay at close to the original.

(We had a wet and dirty application where, in an effort to slow the corrosion and reduce maintenance, the steel plate was replaced with a Type 304 stainless plate. There was a fair amount of sharp sand in the flowing liquid and the result was that the oxide was scraped off the stainless and it failed much faster than the hardened steel.)

In understanding how those local conditions can change the corrosion rates there can be tremendous variations as a result of the corrodent flow rate and some of these can be seen in Figure 7.11 where mild steel in oxygen saturated water (such as would be seen in a bubbling brook) was exposed to a series of flow rates. This is data from a

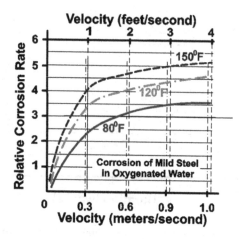

FIGURE 7.11   Corrosion rate variation with flow velocity and temperature.

series of INCO tests and not only shows the effect of the flow scrubbing the samples but also is another example of the effect of increased temperatures.

Another view of the effect of changing flow velocities on the corrosion rate is shown in Figure 7.12, but this also introduces the concept that different metals prevent or reduce corrosion with different mechanisms. The three materials in the example corrode at greatly differing rates and the interesting points are the differences in the basic corrosion rates and how those rates vary with velocity. The mild steel attack varies tremendously as the velocity changes, but the gray cast iron is

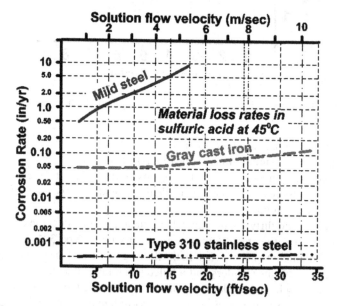

FIGURE 7.12   Showing how the attack rate on three alloys is affected by both solution velocity and metallurgical structure.

much less affected and the 310 SS is almost untouched. The reason for this is that the three materials develop their corrosion resistance by different mechanisms as follows:

- The mild steel actually corrodes very rapidly. It forms a soft ferrous sulfate protective coating, that is easily washed off as the flow rate increases.
- Gray cast iron is a mixture of carbon flakes surrounding the islands of iron and the carbon is extremely corrosion resistant. Here we see it is attacked at less than 1/10th the rate of the mild steel because the free graphite (carbon) tends to protect the surface.
- The 310 stainless steel relies on a durable chrome oxide coating to resist the acid and is almost untouched because the acid can't penetrate that tenacious oxide layer.

THE EFFECT OF ATMOSPHERE AND CONTAMINANTS

In understanding the effect of either solution or airborne contamination on corrosion rates, go back to that corrosion cell shown in Figure 7.3 and think of how the reaction rate could be changed. If the cell surface, either the anode or the cathode, were cleaned regularly the solution would have better access and the corrosion driving force would increase. If the conductivity of the liquid increased the electrons could move more easily and the rate would increase. The effect of contamination is to alter that corrosion rate. Sometime the rate is increased, such as the effect of using salt for road deicing, and sometimes it is decreased such as when a layer of debris reduces the availability of the corrodent.

There are atmospheric corrosion maps of most of the industrial world available and three broad area categories and their corrosion rates for bare mild steel are:

- *Rural*
  - Sub-arctic – 2.5 to 25 µ/yr (0.0001 to 0.001 in/yr)
  - Temperate – 7.5 to 75 µ/yr (0.0003 to 0.003 in/yr)
- *Industrial* – 18 to 38 µ/yr (0.007 to 0.015 in/yr)
- *Marine* – 10 to 60 µ/yr (0.004 to 0.024 in/yr)

Looking at the data for rural areas it is fairly obvious that, because the subarctic areas are drier and colder than the temperate areas, the rates are correspondingly lower. However, the corrosion rates for industrial areas, even though most are in temperate zones, are increased because of the industrial discharges, the sulfates, chlorides, and other chemicals. In marine areas, where there is plenty of moisture available and additional salt contamination, the corrosion rate is even greater.

We already showed data about temperature effects and mentioned that without moisture there can be no corrosion. However, as we all know, the corroding parts don't have to be in a liquid. Some folks have found signs of corrosion at relative humidities as low as 40%, but the general thought is that it can begin at about 60% relative humidity and Figure 7.13 shows how it continually increases until saturation is reached.

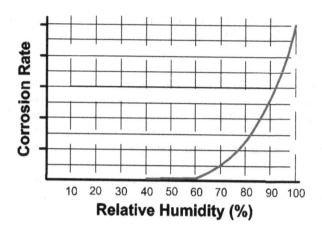

FIGURE 7.13   Effect of relative humidity on steel corrosion rates.

The two specific contaminants that cause the most problems are sulfides and chlorides. An example of the effect of chlorides can be seen by looking at Figure 7.9 and the corrosion rate of Type 304 stainless steel in sulphuric acid. Normally the chrome oxide coating that develops on the stainless is adequate for protection against the acid attack. But if even a small amount of hydrochloric acid is present, the oxide film is attacked and the corrosion rate is increased tenfold or more.

Another example of the confusion and variables involved with material concentrations can be seen by reviewing the effect of concentration on the corrosivity of sulphuric acid in carbon steel. Concentrated 93% sulphuric acid is frequently stored in carbon steel tanks and the tanks can be expected to last for more than 25 years as long as the flow velocity is low. Yet less concentrated sulphuric acid, for example a 50% concentration, would destroy the same tank in a few weeks.

In one job we worked on the personnel specifying the materials of construction on a 5000 Hp centrifugal compressor didn't notice that the process gas was contaminated with about 0.2% hydrogen sulfide. They specified a stainless steel for the rotor material and after several months of operation and repeated vibration problems, we found the corrosion rate was 3.2 mm/yr (1/8 in/yr). The multimillion-dollar rotor was only 6 mm (1/4″) thick to start out with! A change in rotor material reduced the corrosion rate to less than 0.25 mm/yr (0.010 ipy) but it was an expensive lesson.

The *hot wall effect* is an interesting example how solution concentrations can play an important role in corrosion rates. This is a situation where evaporation, combined with poor liquid circulation, concentrates the remaining solution, typically near the tube sheet, the "hot wall", of a heat exchanger. (If there was a liter of an aqueous solution with 1% of a corrodent in the solution and evaporation reduced the total volume to one-half liter, the corrodent concentration would be doubled.) As the evaporation continues the solution becomes proportionally more and more aggressive. The action

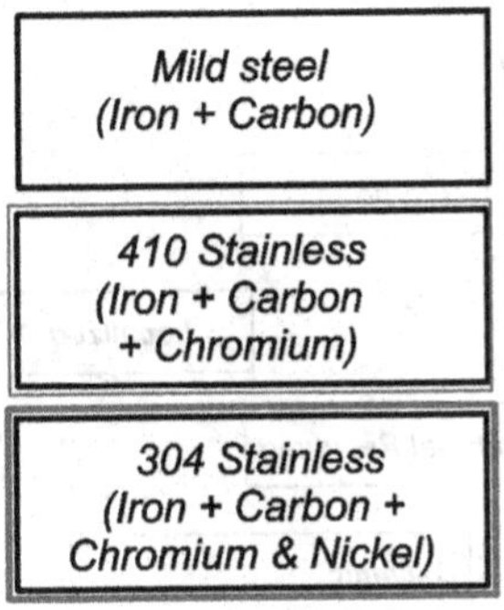

Mild steel (Iron + Carbon)
410 Stainless (Iron + Carbon + Chromium)
304 Stainless (Iron + Carbon + Chromium & Nickel)

FIGURE 7.14   A representation of how alloy additions improve corrosion resistance.

of this *hot wall effect* is commonly seen in heat exchangers but two examples where out-of-service equipment failed as in a similar manner are:

1. A large assembly of Type 304 stainless steel pipes destined for a turbine oil system was hydro-tested using domestic water, then allowed to sit in a field for several months waiting for installation. After the test, the water wasn't completely drained, and the resulting evaporation allowed the chlorides in the solution to become concentrated to the point where the residual fabrication stresses caused stress corrosion cracking.
2. A heat exchanger was taken out of service and cleaned but the tubes were not blown dry. The normal tube sag trapped some solution which, together with the evaporation, resulted in internal pitting of the tubes.

HOW OXIDES PREVENT OR REDUCE CORROSION (ALUMINUM, STAINLESS STEEL, ETC.)

Most metals develop oxides, outer layers where the metal combines with the oxygen in the surroundings, and these oxide "coatings" tend to protect the base metals underneath. For example, iron and steel rust (oxidize) and the rust slows the contact with a corroding material and that slows the corrosion rate. However, the rust doesn't do a really good job of slowing the corrosion and it is fairly easy to remove.

But some metals such as stainless steel and aluminum only takes a second or two to form oxide scales that are tenacious and do a great job at preventing further corrosion.

Figure 7.14 shows the basics of how chemistry creates the oxides that protect stainless steel. Chromium is the basic element that forms the oxide. If more corrosion resistance is needed, nickel is added to the alloy and if even more resistance is needed, molybdenum is added.

THE TYPES OF CORROSION

Depending on who you listen to there are somewhere between eight and eighty forms of corrosion and what we have tried to do is use some of the general groupings and make it understandable for the person who is not a corrosion specialist. For many

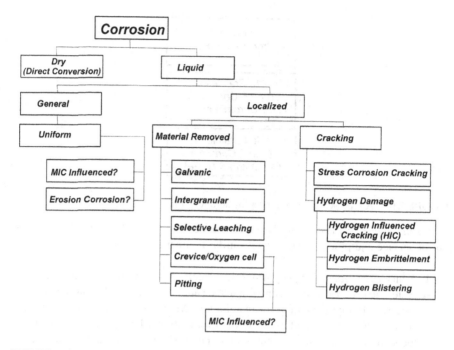

FIGURE 7.15 After many years of looking at corrosion problems in the field, this is how I generally think of them.

years we used the classifications proposed by Dr. Mars Fontana but now feel that a better way to categorize them is as shown below. Then, in the following pages, we'll discuss and show examples of each and explain what can be done to prevent them.

These corrosion mechanisms can be divided into categories as shown in Figure 7.15. Of these mechanisms, only *Uniform Corrosion* results in overall attack of the piece. All of the others are relatively localized, but the underlying mechanisms are similar, despite the differences in appearance.

Unfortunately, corrosion mechanisms frequently work synergistically, i.e., pitting can help cause stress corrosion cracking, etc. This frequently adds another challenge to diagnosing and solving the causes.

Uniform Corrosion

Uniform corrosion is the most common type of corrosion and probably causes 80% of the cost of all corrosion. One of the confusing points about it is that the name *uniform corrosion* refers to the mechanism and not the appearance.

Appearance

This is the rusting (or oxidation) frequently seen on an exposed area of steel. The attack may appear relatively uniform, but more often there will be irregularities caused by local conditions and the surface won't be smooth. Photo 7.1 shows a pair of slip-joint pliers with corroded bands on the face. Referring back to Figure 7.2,

PHOTO 7.1   Uniform corrosion of a forging showing the anode and cathode areas resulting from the internal residual stress.

one can see that it is the anode areas that are rusted while the cathode areas are still clean and like new. Yet we know that, six months from now, the entire pliers set will be rusty. This is a great example of uniform corrosion where today's anode is tomorrow's cathode with the two poles constantly shifting back and forth as the entire piece slowly rusts away.

This is the primary mechanism that causes rusting of cars, bridges, equipment left outside, etc. Photo 7.2 is an example of uniform corrosion that (typically) doesn't look very uniform and shows a railroad bridge support in a large midwestern city. The corrosion damage at the base of the column is the result of the city failing to both paint the column and to clean debris collected around it for a long time. The debris helps retain moisture and is loaded with contaminants that maintain the corrosive attack.

PHOTO 7.2   Long term lack of cleaning and coating and the layer of roadside debris lead to the corrosion of this railroad bridge support.

Of the major forms of corrosion, the penetration rate of uniform corrosion is the easiest to determine and predict. Because of this relatively predictable rate of attack, regular inspections should be able to avoid unexpected failures.

But uniform corrosion is also dependent on flow velocity. Photo 7.3 shows a large heat exchanger with severe erosion in the area between the two tube passes. The plant personnel routinely ignored replacing the baffle gasket, allowing fluid to leak under the baffle plate at high velocity. The flow velocity scrubbed the corrosion product off the tube sheet, keeping the corrosion rate at a relatively high level and corroding that area at a much higher rate than the rest of the tube sheet. After several years, leaks in the tube sheets required replacement at a tremendous cost.

Comments on Uniform Corrosion

1. Trace contaminants, such as sulphur in a rural atmosphere or chlorides in a sea side, can greatly alter the rate of attack. For example, the areas downwind of major airports tend to suffer greater corrosion rates than the upwind areas.
2. Increases in mechanical stresses can more than double the corrosion rate.

Prevention

Control usually involves either removing the corrodent or the moisture. This could involve chemically preventing oxygen and water from getting to the surface and both paints and coatings slow the rate at which moisture gets to the metal surface.

PHOTO 7.3   The corroded line up the center of this tube sheet photo is a groove resulting from the increased fluid velocity from leakage under a tube sheet baffle.

PHOTO 7.4 Despite the nonuniform appearance, this is uniform corrosion of a stairway in a chemical plant and the holes are where the repeated blows from workers shoes damage the poorly applied coating.

A caution about painting. One of the basic rules of corrosion is that the new material is always the anode. So, if a section of a pipe, or a new panel in a steel tank is replaced, these new portions become anodes and tend to corrode more rapidly. There is a temptation to paint those new sections but, if there are any holidays (holes) in the coatings, the corrosion currents will be concentrated at the holidays and the attack rate will be amazing. If you are going to use a coating, either coat both parts, old and new, or just coat the cathode (Photo 7.4).

Fretting

Fretting is a form of uniform corrosion that occurs when tiny (less than 0.001″) movements between two parts cause a microscopic welding and tearing of materials.

Appearance

Dark brown or black discoloration between two ferrous parts. (Black is usually an indication that water was present.) Photo 7.5 shows the impeller fit on a pump shaft and the brownish-black discoloration is fretting, the result of looseness between the two components.

With stainless steels fretting often causes a surface that initially looks polished but is actually slightly pitted and not as smooth as it was.

PHOTO 7.5   The discolored potion of this shaft fit is fretting. The brownish-black debris is oxidized steel and results from movement between the two components.

Comments

1. Oftentimes seen between a bearing ring and the shaft or housing, between a hub and a shaft, and between the strands in a wire rope.
2. False brinelling in bearings is a form of fretting.
3. Fretting will substantially reduce the fatigue strength of the fretted parts.

Prevention

Frequently difficult and depends on preventing movement between the parts. This can be done with tighter fits or adhesives or by changing the design.

GALVANIC CORROSION

Galvanic corrosion is probably the second most common mechanism of corrosion. It happens when two metals or alloys with differing electrical properties (EMFs) are physically connected and immersed in a conductive liquid. They develop an electrical potential and when the current is allowed to flow, the more active metal (the anode) will be consumed as part of the electrochemical reaction.

Figure 7.16 shows a galvanic series in seawater at ambient temperatures. Looking at it, the metals on the top of the scale, those with the most negative voltages, are more active than those below. So whenever two metals are connected, the more active one corrodes and protects the lower one.

A good example of this is the galvanized (zinc) coating frequently used on steel pipe. The zinc is anodic to the steel it is plated on and protects the steel from corrosion. In a similar manner, the zinc anodes mounted on the wetted area of outboard motors protect the aluminum lower units from corrosion.

Photo 7.6 shows an example of galvanic corrosion. The pieces were taken out of a water well nine months after the well had been repiped. On the left is a bronze foot

Galvanic Series in Sea Water

Magnesium
Zinc
Galvanizing on Steel or Iron
Aluminum (5052, 3004, 3003, 1100, 6053)
Cadmium
Aluminum (2000 series)
Mild Steel
Type 410 Stainless Steel (active)
Type 304 and 316 Stainless Steel (active)
Lead
Naval Brass
Yellow Brass
Aluminum Bronze
Copper
70-30 Copper Nickel
Nickel 200
Monel 400
Type 304 Stainless Steel (passive)
Type 316 Stainless Steel (passive)
Silver
Titanium
Graphite
Platinum

FIGURE 7.16   This shows the Galvanic Series in seawater. Seawater is used because there is a universal standard and variations in water chemistry and temperature can change the relative alloy rankings.

valve that is in like-new condition. To the right of it is a pipe nipple that was originally galvanized. (In the middle of the galvanized pipe is a shiny section that is a piece of cellophane tape wrapped around the pipe.) The nipple had been in the well for many years while the galvanizing was doing its duty protecting the steel. However, when the well was repiped the fitter shortened the nipple and recut the threads on it, exposing clean steel. Unfortunately, after years of use, the galvanizing was no longer there with most of the zinc having corroded to zinc hydroxide, leaving the pipe

PHOTO 7.6   A well pipe assembly made from bronze, steel, and what was originally a galvanized pipe. But the zinc had long ago corroded to zinc hydroxide and couldn't protect the freshly cut steel threads.

unprotected. As a result, the assembly surfaces exposed to the water consisted of bronze, freshly cut steel, and old steel. The resulting galvanic corrosion is obvious.

Appearance

Recognition is relatively easy because the corrosion of the more active metal starts right at the junction of the metals, then decreases with distance.

Comments

1. The galvanic series dictates which metal is attacked. The larger the potential difference (voltage) between the metals the greater the rate of corrosion and the actual rate of metal loss can be predicted based on the current (amperage) flow.
2. Temperature and environment can affect the corrosion rate.
3. The relative sizes of the cathode and anode are critical. A large cathode will generate a large current and concentrating the current on a small anode will result in rapid loss of material and penetration. This is the *area effect*. (See *Prevention*, below.)
4. Whenever new materials are installed in a corrosive environment, the new materials are always anodic to the older ones. (In one installation we worked on there were two pipelines with identical metallurgies, side-by-side. The newer one had five times as many leaks as the older one.)
5. To avoid galvanic attack, weld filler metal should have as much or more alloy content than the base metal.
6. If there are two metals that are electrically connected and the plan is to coat only one of them, be sure it's the cathode. (There are always holidays in coatings. If the anode is coated, the corrosion current generated by the cathode will concentrate at a holiday in the anode coating and result in rapid perforation.)

Prevention

Usually involves insulating against contact between dissimilar metals. For example, when a new section of buried pipe is installed an insulating flange kit is installed. This involves using gaskets to separate between the pipe flanges and insulating bushings and insulating washers to ensure there is no conductivity between the two pieces.

However, another approach is to make the anode much larger than the cathode and the *area effect* is very important in galvanic corrosion. The first galvanic corrosion that we could find reference to occurred in the British navy in the mid-1700s. At that time the wooden frigate *Alarm* was fitted with copper sheathing to protect the bottom from attack by marine worms. The sheathing was held in position with iron nails, the inevitable galvanic cell was developed, and the nails corroded rapidly. (When the ship returned to port with the copper missing the entire crew was jailed under suspicion of theft and it took some time before they were freed.) This is an excellent example of the area effect where the large cathode generated a current that was applied to the much smaller anode and the loss of material had a rapid effect. If the areas were reversed, with a large anode and a small cathode, not only would the corrosion current be less (and a lesser amount of material removed), but

PHOTOS 7.7 AND 7.8   The first photo is of a small galvanized steel nipple. It was screwed into the cast iron pump housing shown in th e second photo. The top of the housing was always damp and the protection of the galvanizing was eventually exhausted. The repair consisted of replacing the steel nipple with a copper one. Now the cast iron housing will be the sacrificial one, but the damage will be distributed over a much larger area and it will be many years before a replacement is needed.

also it would be applied over a larger area with a correspondingly lower overall penetration rate.

Photos 7.7 and 7.8 show another example of using the area effect to solve a galvanic corrosion problem. In the first view we see a galvanized steel nipple where the galvanizing has been exhausted and a hole developed. This nipple was mounted on top of a cast iron well pump that was constantly moist from condensation and the steel pipe corroded through while it protected the cast iron housing. The repair consisted of replacing the steel nipple with a copper fitting. Now the cast iron housing will be the anode but the corrosion will be so distributed that we will all be long gone before the housing needs replacement.

Probably the most expensive example of galvanic corrosion we were involved with involved a compressor impeller that was coated with flame-sprayed tungsten carbide. (The cost, not including production losses, would be about $2,000,000 in 2019 dollars.) There were holidays in the coating and galvanic corrosion occurred at the interface, flaking off large areas of the coating and unbalancing the rotor about every two weeks!

Selective Leaching

Selective leaching is a form of galvanic corrosion where one component of a metal that is not a true alloy is attacked, leaving the other in place. The most common examples of it are:

- "Graphitic corrosion", typically on buried cast irons pipes, where the iron is removed and the graphite left behind,
- Dezincification of zinc-copper alloys.

Appearance

The surprising part of selective leaching examples is that they look almost unchanged from their original appearance, the difference being that one of the components of the material is no longer there. (It has been dissolved out of the solution.)

Photo 7.9 shows a 20 cm (8″) cast iron fire water line that had been buried for 30 years when it sprung a leak. When the plant personnel dug up the section the crack was obvious, but the rest of the pipe looked like new.

Comments

1. The term "graphitic corrosion" is technically incorrect but it is commonly used to describe the selective leaching of cast iron pieces.
2. This attack mode is common in buried cast iron sewer and water pipes. When a section of buried pipe is uncovered it looks as though there is no damage, but the surface of the pipe can be carved with a sharp knife because the iron is no longer present, and this part of the pipe wall is essentially graphite. (Cast iron is essentially iron in a graphite matrix and the graphite is the cathode. When very slow moving water is present the frequent result is the galvanic attack of the iron and what remains is a graphite pipe.)

Prevention

Selective leaching in buried piping can be prevented with cathodic protection and with coatings, otherwise it requires a materials change.

PHOTO 7.9 A fire water pipe that lasted about 30 years. When it was removed from service the leak was visible but it wasn't until the piece was sandblasted to remove the graphite that the extent of the attack became apparent.

INTERGRANULAR CORROSION

Intergranular corrosion is occasionally found in stainless steels but also shows up in other corrosion resistant alloys. It is caused by changes in the metal's chemistry and is usually the result of welding or improper heat-treating procedures.

Appearance

If welding causes it, the most common location for it is the area immediately next to the weld. In other cases, errors in heat treatment can affect entire pieces and portions of the metal disintegrate and crumble. When welding causes the problem, it will look like a groove has been machined alongside the weld but not contacting it. Otherwise, the surface will appear rough and feel "sugary".

Very early in my career I was the manufacturing engineer in a plant with a pickling operation. Steel parts were put in a large welded Type 316 stainless steel basket, washed, and then dipped in an acid solution. The baskets constantly corroded in grooves similar to that shown on the right-hand side of Figure 7.17 and the grooves in the 6 mm (1/4″) thick stainless were about 3 mm (1/8″) from the weld. It was a routine maintenance procedure, about every six months, to pad weld the grooves, but they continually reappeared. Then we asked the corporate metallurgy department for some assistance and they suggested we make new baskets from Type 317L stainless. The Type 317L baskets lasted a great many years with almost no corrosion.

Comments

1. The mechanism of how it occurs with weldments is as follows:
   - The stainless alloy contains both carbon and chromium.
   - When the alloy is heated above about 525°C (1000°F), the carbon becomes mobile and tends to unite with the chromium at the grain boundaries and form chromium carbides.
   - The result is that there is less chromium available to resist corrosion in the areas immediately adjacent to the grain boundary. The chrome-poor section around the boundary then becomes anodic to the rest of the grain, is corroded, and the individual grains fall out.

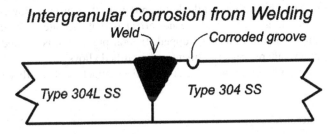

FIGURE 7.17 An example of how the low carbon stainless steel alloys are not as severely attacked as the higher carbon versions.

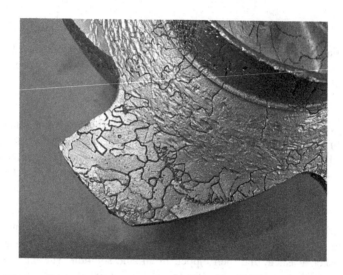

PHOTO 7.10 This is a stainless steel pump impeller that was improperly heat treated. When exposed to concentrated phosphoric acid the grain boundaries were attacked and the grain structure can be clearly seen in the photo.

2. In the past, selective leaching commonly happened with Type 304 or Type 316 stainless steel exposed to corrosive liquids. These have been largely replaced with the low carbon grades, 304L and 316L, which have less carbon available to "steal" the chromium.
3. The "L" designation refers to the fact that the amount of carbon is limited. (Type 304 allows as much as 0.08% while 304L limits the carbon content to 0.03%.) As a result, fewer chromium carbides are formed at the grain boundaries and the rate of attack is greatly reduced.
4. The downside to the reduced carbon level is that the "L" grades are somewhat weaker than the higher carbon alloys, but some of them have other alloying materials, such as nitrogen, to increase their strength.
5. There have been instances of intergranular corrosion of aluminum alloys and brasses.

Prevention

With weldments, the most common approach is to use low carbon material (304L, 316L) so there won't be carbon available to unite with the chrome. However, there is still some chrome available and the net effect is that the rate of intergranular corrosion is greatly slowed but not stopped. If total prevention is needed the best approaches involve altered heat treatment and modified alloys.

In some cases, intergranular corrosion involves attack such as that shown in Photo 7.10.

EROSION CORROSION

As mentioned earlier in this chapter, as the flow rate of a corrodent increases the corrosion rate almost always also increases. This is called flow accelerated corrosion (FAC).

However, the difference between FAC and erosion corrosion is that FAC tends to scrub the surface clean and expose it to fresh corrodent while erosion corrosion is a result of a combination of high velocity mechanical removal of an oxide scale and corrosion, thereby greatly increasing the loss rate. Erosion corrosion frequently happens when there are solids in the liquid or liquid particles in a high velocity gas flowstream.

Appearance

Smooth bottomed pits, frequently either horseshoe shaped or superimposed on one another, with a directional pattern related to path of fluid. The top edge of the pit is frequently sharp and pointed downstream. Photo 7.11 shows a pair of 2.5 cm tubes that have been removed from an ammonia vaporizer and cut in half.

Another example of erosion corrosion is shown in Photo 7.12 where a piece of a pump impeller shows a very definite flow pattern from the left to the upper right.

Comments

Erosion corrosion thrives on high velocity conditions and occurs in ells, tees, valves, pump cases, impellers and other places with high velocity and turbulence.

For erosion corrosion to take place there has to be a corrosion mechanism at work.

Prevention

Reducing the flow rate, changing to a harder material, or changing to a more corrosion resistant material. In cases where there are solids in the flow stream, eliminate the solids or reduce their size.

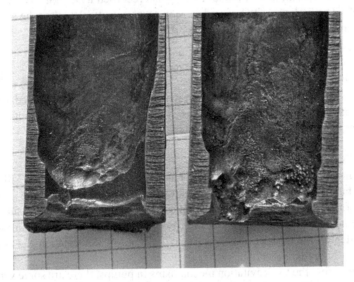

PHOTO 7.11 Two corroded tubes where the liquid ammonia comes in the bottom of the tubes and flashes into a gas. There is some deformation on the lower edges of the tubes from the removal tools but just above that can be seen an area where the wall has been substantially thinned. The tube to the left shows more classical symptoms of erosion while the tube to the right shows obvious erosion and corrosion attack.

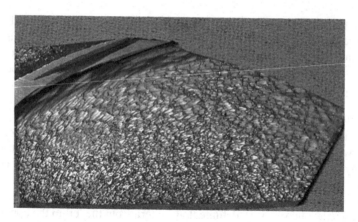

PHOTO 7.12   An eroded pump impeller.

The most amazing erosion corrosion we ever saw was on a boiler tube in a sulphur burning boiler. The plant had been having problems with tube failures and we had just completed an extensive series of thickness tests on the tubes. We had found that some were thin, and the plant put the boiler back on line with the idea that they would change them in the near future. Eventually, one of the thin tubes failed and blew hot water and steam into the burning sulphur stream, washing across the adjacent tubes. Then a second tube failed, but it wasn't one of the thin tubes and we had recorded its thickness at 5.1 mm (0.20″). The high velocity hot acid erosion corrosion attacked the tube at a rate of approximately 1780 meters/yr. (584 ft/yr.)!

Cavitation

Cavitation is a form of erosion corrosion where vapor bubbles are formed, and their collapse causes tremendous impact forces. It is usually found in pumps but can occur anywhere that the forces cause vapor bubble development and collapse.

Appearance

It frequently looks similar to fine pitting. On pumps, as shown in Figure 7.18, starvation cavitation damage occurs on the back (negative pressure) side of the vanes near the eye of the impeller while recirculation cavitation appears on the pressure side of the vane near the OD. Occasionally microscopic or metallographic examination is used to identify the damage and the surface grain plastic deformation can be plainly seen under high magnification.

In comparing the two cavitation mechanisms in pumps, there are some very obvious differences. The figure shows how the locations differ but in addition to that:

- *Starvation cavitation* is caused by a lack of pressure on the fluid at the eye of the impeller. The fluid vaporizes then the vapor bubbles collapse as they progress into higher pressure areas. The result is a noisy pump that sounds

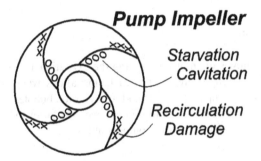

FIGURE 7.18 A drawing of a pump impeller showing the difference in the locations of the cavitation mechanisms. Starvation cavitation results from an inadequate liquid supply while recirculation happens when the discharge is restricted.

like it is pumping nuts and bolts and the forces are tremendous and can destroy the pump bearings in less than a day.

- *Recirculation cavitation* is caused by the pump's inability to properly discharge the fluid. When the discharge is throttled, as the vane passage passes the high-pressure area of the volute some of the liquid is forced back down the passage. In severe cases recirculation cavitation can destroy an impeller in less than a year.

Photo 7.13 shows a view of a pump impeller with holes alongside the passage and at the end of the vane that were caused by recirculation cavitation.

Comments

1. Cavitation forces can be as high as 80,000 psi.
2. Austenitic stainless steels and aluminum alloys are particularly vulnerable at high flow velocities.

PHOTO 7.13 Recirculation cavitation damage on a pump impeller.

Prevention

Three approaches are:

a. Improve the flow stream by eliminating the cavitation action.
b. Select higher strength alloys or hard facing alloys where the surface is stronger than the impact forces from the cavitation bubbles.
c. Use corrosion resistant alloys to remove one portion of the erosion-corrosion action and greatly improve component life.

> In one cavitation example there was a ~190 kW (250 hp) 1800 rpm pump and recirculation cavitation absolutely destroyed the impeller in less than a year. It pumped water from a local stream that was used by a plant for cooling. The impeller was changed to a hardened high chrome cast iron and an inspection after six months of operation found the original machining marks were still visible. The pump was shut down ten years later and the impeller was still in good condition.

CONCENTRATION CELL CORROSION

Concentration cell corrosion is caused by differences in concentration in the solution contacting the target material. These concentration differences cause electrical potential differences and a corrosion cell begins.

Appearance

This relatively common form of corrosion is often seen at the "waterline" of a liquid and on buried materials. Photo 7.14 shows four structural steel bolts where the concentration cells at the bolt heads started the corrosion that has begun to destroy the surrounding paint.

Comments

1. Similar conditions cause oxygen differential cell formation under deposits or in other areas deprived of oxygen, i.e., between dirt, corrosion products, or scale and metal wall.
2. The difference in available oxygen (an oxygen differential cell) contributes to most buried pipe leaks occurring on the bottom of the pipe.

Prevention

Combating it can be accomplished by changing materials but a common approach on tanks and pipelines is cathodic protection.

Crevice Corrosion

Crevice corrosion is a very localized concentration cell mechanism that occurs in crevices wide enough to allow liquid to enter, but so narrow the liquid is stagnant. (Cracks of about 0.1 mm [0.004″] are just right!)

PHOTO 7.14   The corrosion raising the paint around these four bolts is concentration cell corrosion and is the result of the chemistry plus problems with doing a good job of coating.

Appearance

Location is the key – under rivet heads, bolt heads, nuts, lap joints, incomplete weld joints, area between gasket and flange faces, absorbent gaskets, etc. The attack shown in Photo 7.15 started at the surface of this Type 304 stainless steel instrument tube that was immediately adjacent to another piece. When the adjacent piece was removed the attack was obvious.

Comments

1. Most often occurs in fluids containing chlorides.
2. Austenitic stainless steels are particularly susceptible.

PHOTO 7.15   Crevice corrosion of a stainless steel tube that was tight up against another tube.

Prevention

Eliminate crevices and deposits, change material, and apply coatings. (This sounds like a good correction practice, but it is *much* easier said than done.)

PITTING CORROSION

Frequently appears as localized, small, almost random holes in an otherwise unaffected area. It usually occurs in materials that have a protective coating. Holes can be deep or even through the wall. Pitting frequently acts as a stress concentration factor, starting fatigue cracks, and is sometimes difficult to detect as the pits frequently start under deposits and are covered by the corrosion products.

Appearance

There are two general types of pits:

1. In carbon steels they are wide and shallow and progress at a very slow rate. Photo 7.16 shows some pits in a mild steel hydraulic elevator cylinder. The cylinder was buried in a wet soil and wasn't properly coated and wrapped so the water leaked in around the weld.
2. In stainless steels the pits often grow bigger as they progress through the part and the entrance may even be minute.

Photo 7.17 shows two aluminum housing covers that were separated by about 20 cm. The atmosphere around the housings was very mildly corrosive to aluminum and the plant had put a condensate hose on the left cover to prevent the product from freezing in the housing. The increase in temperature resulting in a good example of Arrhenius' rule and Photo 7.18 is a close-up of the heavily pitted surface.

Comments

1. Pitting occurs commonly in austenitic stainless steels, but also is found in aluminum and carbon steel (and other metals that rely on protective oxide coatings), almost always in stagnant areas that allow the localized corrosion

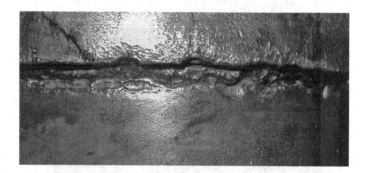

PHOTO 7.16   Pitting corrosion of a steel tube that started alongside the weld because the residual stress from the welding made the adjacent parent metal anodic.

PHOTOS 7.17 AND 7.18 The attack on the left casting was much more severe because it was heated by the condensate.

cell to develop. The presence of a liquid is a necessity and we have seen it happen at temperatures over 600°C (1100° F).

2. Cause – Local cell develops and breaks through the protective (oxide) film. In stainless steels the combined presence of soil deposits and chlorides frequently accelerates the process. Photo 7.19 shows the tube sheet of a Type 316 stainless heat exchanger.

   One serious problem lies in the fact that pitting is very difficult to stop because the inside of the pit is anodic to the surroundings and oxygen can't readily enter the pit to reestablish the oxide coating. In this example they ended up with holes in several of the tubes.

3. Pitting can be a rapid process. 0.5 mm (0.02″) per month is not uncommon in stainless steels.

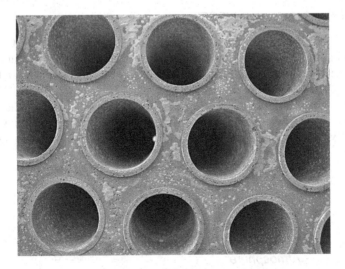

PHOTO 7.19 Pitting of the end of a heat exchanger that happened when the plant wanted to accelerate the cleaning of the heat exchanger bundle. They intentionally increased the chloride concentration in the cleaning solution and perforated several tubes.

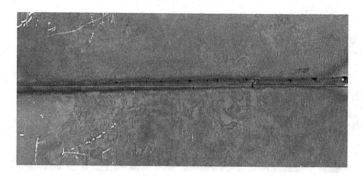

PHOTO 7.20   Pitting alongside the weld in a Type 316 stainless steel duct. A good practice is to always ensure that the weld metal is more corrosion resistant than the parent metal.

Prevention

1. Use of pitting resistant metals. In piping and tanks increased flow rates and/or frequent cleaning can prevent deposits from forming and can prevent chlorides from concentrating, both of which help prevent pitting from starting.
2. As mentioned above, correction is almost impossible. Once a pit starts the anodic action within the pit interior acts to continue the pitting process.
3. Photo 7.20 shows pitting along the edges of a stainless steel weld in a stainless duct. The duct carried acid gases that were occasionally condensing. The weld metal was ostensibly the same as the duct alloy, but in reality it had a little less chrome – so it became anodic. The oxide film prevented uniform attack but couldn't stop the pitting in the highly stressed areas.

Stress Corrosion Cracking (SCC)

Almost every presentation on SCC includes a discussion on the "SCC triangle" shown in Figure 7.19. As shown in the diagram, in order for SCC to occur there has to be sufficient stress, the material has to be sensitive to the atmosphere, and the correct atmosphere has to be present.

An excellent example of the need for all three can be seen in a milk tank inspection project we worked on. For many years, we routinely inspected the welds in a series of Type 304 stainless steel tanks (shown in Photo 7.21) to ensure there were

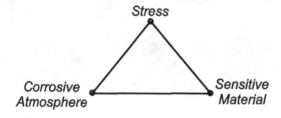

FIGURE 7.19   Stress corrosion cracking has many variations, but they all require these three components.

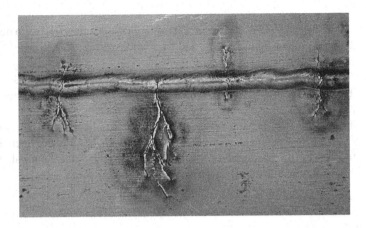

PHOTO 7.21   Stress corrosion cracking in a 40+-year-old tank that appeared shortly after they raised the temperature of the cleaning solution.

no cracks, and for many years we found nothing of concern. Then one year we found a small crack in one of the welds. The plant people said nothing had changed, but the next year we found several cracks and had lengthy meetings to determine why they had appeared. Eventually it came out that they had increased the temperature of the rinse water used to clean the tanks. The atmosphere where the cracking started hadn't changed and the metal certainly hadn't changed, but the increased temperature increased the stress in the welds – and SCC appeared.

Appearance

These are ragged irregular fine cracks with multiple branches while the rest of the piece looks absolutely unaffected. The cracks frequently start at pits. Photo 7.22 shows a 1.2 cm (1/2″) tube with a lengthy jagged and disjointed stress corrosion crack.

> The need for the stress is interesting. For about 15 years we inspected a series of large process vessels every six months. When we started, we knew there were stress corrosion cracks in them and our job was basically to monitor the crack size and number to ensure there was no chance of a catastrophic failure.

PHOTO 7.22   These branched irregular cracks are typical of stress corrosion cracking.

While doing these inspections we started using ultrasonic flaw testing to determine the true extent of the cracking and we found:

- The cracks were in the area affected by the residual weld stresses.
- They were growing deeper through the welded stainless steel, but not longer.
- They were low pressure vessels operating at only about 5 psi (about 1/3rd atm). The plant went through several slight pressure increases to increase the production rate and every time they did, the number of cracks jumped up shortly afterward.

Comments

1. The amount of stress can be small, as low as 14 MPa (2000 psi), but it can also range up to almost the yield strength of the material.
2. Discontinuities such as pits, gouges, and deep scratches are frequently starting points. Photo 7.23 shows a view of a small diameter instrument tube as seen through a microscope. The stress corrosion crack is obvious, and a close examination can see the pits where it started.
3. It can sometimes be found at ambient temperatures, but it is more likely at higher temperatures. We have seen it in equipment that never saw temperatures above 25°C (75°F) but it becomes much more common above 50°C (122°F). Unfortunately, there are people (and books) that say it does not occur below 50°C (122°F).
4. Troublesome environments – Caustic soda for Inconel®, Monel®, and nickel; chloride and fluoride solutions for austenitic stainless steels; caustic and nitrate solutions for carbon steel; ammonia compounds with copper alloys. All metals have some condition that causes SCC.

PHOTO 7.23 These stress corrosion cracks started at the pits. As their diagonal layout on the tube suggests, there is a torsional stress in it.

5. "Caustic embrittlement" is a form of SCC that happens when carbon steels are exposed to caustics, and it generally occurs at elevated (over 50°C/125°F) temperatures but can happen at much lower temps.

Prevention and Correction

A. Prevention – Remove the stress, change the atmosphere or the material. (Sometimes decreasing temperature can help.)

B. Once the material has cracked it is almost impossible to stop further cracking. This is a corrosion mechanism so elimination of all moisture should effectively stop the crack growth progression but that is almost impossible to accomplish. However, if the stress causing the cracking decreases, the cracking will stop. (For example, if the residual welding stress is the source of the stress, the cracking will decrease as the field stress decreases.)

C. One serious limitation of materials that have visible SCC is that there are many more microscopic cracks. The SCC cracks will grow in a predictable manner, but the microscopic cracks will have seriously reduced the toughness of the material. (Stress corrosion cracked components can be used but it is important that the limitations be recognized and cautions applied.)

HYDROGEN DAMAGE

This description covers a multitude of problems and is described in sometimes confusing terms. Understanding the hydrogen damage mechanisms starts with the basic corrosion mechanism. Atomic hydrogen is generated at the cathode and most of the hydrogen atoms join with other atoms to form the familiar molecular hydrogen gas, H_2. But some of the atoms remain unattached and they are much smaller than the voids between the iron and steel atoms. These hydrogen atoms enter the metal and travel through it causing a variety of problems including hydrogen embrittlement, hydrogen influenced cracking (HIC), and hydrogen blistering.*

Hydrogen Blistering

This is the easiest to understand of all the mechanisms. When free hydrogen migrates through a metal and finds another hydrogen atom the two atoms unite to form molecular hydrogen, H_2. The unifying forces that are developed when the two form this molecule are much stronger than the metal and a blister develops inside the metal wall (Photo 7.24).

Appearance

These blisters range in size from that of a small pea to the size of a football. It frequently happens at elevated temperatures and pressures where there is substantial dissociation of the hydrogen molecules but can occur at ambient temperatures.

---

* The hydrogen damage mechanisms are complex and probably the best text for a failure analyst is *Hydrogen Embrittlement of Metals: A Primer for the Failure Analyst* by M. R. Louthan Jr. and it is available at https://sti.srs.gov/fulltext/WSRC-STI-2008-00062.pdf.

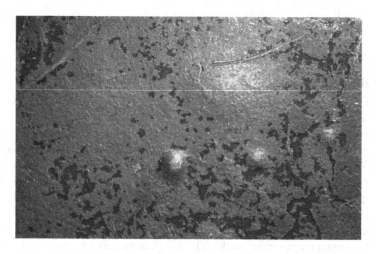

PHOTO 7.24 A steel plate with some hydrogen blisters that are about 9 mm (3/8″) in diameter.

Prevention

Correction is difficult and action is most often directed toward monitoring. However, inhibitors and coatings have been used successfully.

Comment

Most of these occur during hydrocarbon processing but large carbon steel sulphuric acid storages are also a common location for hydrogen blisters. In the acid storages the hydrogen is generated by the reaction that forms the ferrous sulfate layer and tends to collect in areas along the center of the plates. Any welding or torch cutting of these acid tanks and lines is frequently accompanied by sputtering and popping of many small blisters.

Hydrogen Influenced Cracking (HIC)

Current thinking has the mechanism as:

- Hydrogen migrates through a material and unites with other materials to form hard brittle materials.
- The metal then loses ductility in the area of these intermetallics and elevated pressures are present.
- Cracking then can occur at loads well below the tensile strength.

It is commonly seen in higher strength (HRC 32 and above) steels near processes that liberate hydrogen but can occur in lower strength steels in more aggressive atmospheres.

Appearance

The appearance is essentially indistinguishable from SCC with ragged disjointed cracks as seen in Photo 7.25. This is a cross-section of an ASTM A 193 Grade B7

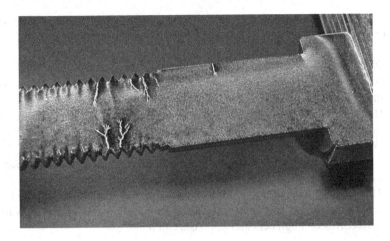

PHOTO 7.25 The fact that these cracks are so disjointed indicates that the cause has to be internal.

bolt that was exposed to a mildly corrosive atmosphere. The bolt is being wet fluorescent magnetic particle tested and, using a black light, the cracks show up as white lines.

This mechanism can also appear as seen in Photo 7.26. This is a spring with a corroded mounting surface. The end coil was ground flat and it rested on a plate that was frequently wet. Eventually some corrosion occurred, and the generated hydrogen combined with the high hardness of the spring resulted in several failures.

PHOTO 7.26 A cracked spring that started with corrosion and hydrogen damage.

This is a photo made through a 7X microscope and the semi-circular original crack is plainly visible.

Comments

1. Any process that generates hydrogen, i.e., corrosion, plating, welding, and other industrial processes, can cause these problems.
2. Higher strength steels are much more sensitive to the presence of atomic hydrogen.
3. With welding, this is the reason why low hydrogen rods are beneficial.
4. Appearance is typically a jagged crack with fine branching cracks adjacent to the primary one (similar to SCC).
5. There is confusing terminology about what we have called HIC. In the fastener industry this mechanism is known as *hydrogen embrittlement* while most corrosion engineers and scientists call it HIC.

Prevention

- In manufacturing processes, if the hydrogen source cannot be eliminated, the parts are baked at 190°C (375°F) for 23 hours to drive out the hydrogen. (In order to prevent permanent internal damage to the parts, this has to be done in a relatively short time after the hydrogen is absorbed.)
- In plant operations involving corrosion the materials have to be changed to eliminate the problem.

Hydrogen Embrittlement

In the eyes of corrosion engineers and scientists, there are two forms of hydrogen embrittlement:

1. The metal absorbs hydrogen and is converted to a brittle non-metallic compound. For example, Photo 7.27 shows a piece of titanium tubing that is about 2 cm (1 in) in diameter. Looking at the right side of the photo it can

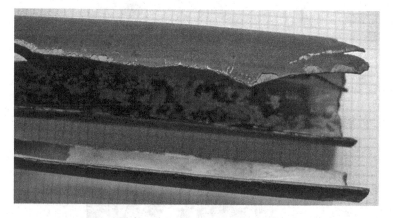

PHOTO 7.27   True hydrogen embrittlement where the ductile titanium has absorbed hydrogen and become the very brittle titanium hydride.

be seen that there are chips out of the tube that appear to be classical brittle fractures. Yet the portion of the tube that was not in the gas stream is still as ductile as a paper clip! The tube was exposed to steam and the free hydrogen atoms have converted the external portion of it to titanium hydride and it is almost a brittle as glass.

2. The other embrittlement mechanism involves the hydrogen atoms meeting scattered nonmetallic ions within the metal and forming compounds that result in internal pressures that greatly reduce the tensile strength of the metal.

A SPECIAL CATEGORY: MICROBIOLOGICALLY INFLUENCED CORROSION

Microbiologically influenced corrosion (MIC) deserves special mention because of its effect in causing corrosion. Figure 7.10 shows the eight corrosion categories and three of them, uniform corrosion, pitting, and crevice corrosion are sometimes affected by microbial action. Twenty-five years ago, many corrosion specialists thought that perhaps five or even 10% of the corrosion was caused by MIC. Today the consensus is that a substantial percentage, perhaps as much as 40% of all corrosion, is influenced by these living organisms. The organisms don't actually do the corroding, they act to accelerate the corrosion. (A good example of this was the brine lines that ran from our brine wells to the plant. There were microbes that lived in colonies along the pipe wall. The microbes consumed the sulphur compounds that were in the brine and secreted sulphuric acid, which didn't help the pipe at all.)

The situations range from deposits caused by slimes that tend to be resident sites for crevice corrosion to microbe colonies that physically consume metals. There are a multitude of possible situations and testing is recommended if MIC is suspected at all.

- Identification is difficult because microbes are frequently present but may not be involved with the corrosion. Three sources of test kits used to detect the presence of MIC are Bioindustrial Technologies Inc., DuPont, and Hoch.
- Environment – Liquids must be present. Observed temperatures range from 0 to 110°C (32 to 230°F). Observed ph range is from 0 to 14. Normally it does not occur in flows greater than 1.5 meter/sec (5 ft/sec.)
- Combating MIC depends on the specifics of the case. Alternatives include biocides, process changes and material changes.

BIBLIOGRAPHY

ASM Handbook of Corrosion Data, Edited by Bruce Craig, 1989, ISBN: # 0-87170-361-0.

ASM Handbook, Volume 13, ASM International, Metals Park, OH, 1987, ISBN: # 0- 87170-007.0.

ASM Handbook, Volume 13A, ASM International, Metals Park, OH, 2003, ISBN: # 0- 87170-705-5.

The Effects of Velocity on Corrosion, International Nickel Company, New York, 1960

Fontana, Mars and Greene, Norbert, *Corrosion Engineering* (3rd edition), McGraw Hill, New York, 1979, ISBN: 978-0070214613.

8 Lubrication and Wear

In understanding the effects of various lubricants and how they affect machine life, it makes good sense to first look at some simplified versions of how the basic bearing types work, then review how lubricants are formed and how their additives work, then return for a more detailed look at various applications and the critical nature of the lubricant interactions.

Before we do that, one very important point is that there is much more to the typical lubricant than what meets the eye. I've heard far too many uneducated people say something like "Oil is oil" and not realize what they are using may have 20% additives or it may have only 1% additives. They also don't understand that the tube of grease in their hand may be 75% oil and 8% additives and be designed to have the oil slowly separate from the thickener, or maybe the oil and thickener are designed to never separate. Lubricants start with a base oil and then the additives help them do their magic.

A story that illustrates those differences happened while I was still working in the chemical plant. I'd noticed that one of the cars in the engine room operator's parking lot smoked whenever it was started, and that was really unusual because the car was only three or four years old. I noticed it a couple of times and it was really just a curiosity. But one day we were working on a job in the engine room and were waiting, and waiting, and waiting for the crane. (These were big, old steam engines. Even a connecting rod bolt was over 40 kg [100 pounds] and almost everything was lifted with the crane.) While we were waiting and gabbing, I asked the crew about the car that smoked, and they all laughed. One of them said, "Oh, that's Crazy George" and went onto explain that George thought he was "putting one over on the company" by taking a quart of engine oil every week and using it in his car. He was stealing steam engine oil, with about 0.5% additives, and putting it in his car engine instead of the API graded oil that that typically had about 17% additives! Education may be expensive, but it's far less expensive than making mistakes.

THREE TYPES OF LUBRICATED CONTACT

There are three basic types of lubricated contacts that are generally seen in industry:

1. Rolling element (ball and roller) bearings such as those used in most common motors, reducers, and pumps
2. Plain (sleeve) bearings in common use with large motors, generators, and the main and rod bearings in most of our automobile engines
3. Sliding bearings where pieces are in contact with and sliding over each other, such as hydraulic cylinders, trailer hitches, and slider plates

177

In looking at these bearings, they involve three very different lubrication regimes that require different oils and greases to perform their jobs most efficiently over time.

LAMBDA – THE LUBRICANT FILM THICKNESS

Viscosity is a measurement of how readily the liquid lubricant flows when subjected to a load. High viscosity fluids, such as tar, tend to resist and stay in position while low viscosity fluids, such as water, are easily displaced.

Probably the most important lubrication concept to understand is that of the λ ratio, the relative film thickness. It typically ranges from 0 to 4 and is the result of the lubricant viscosity, the difference in speed between the two operating pieces, and the roughness and shape of the mating surfaces. In Figure 8.1, the right half of the chart is largely controlled by the lubricant's viscosity while the left side involves a complex relationship between the oil, any grease thickener, and the additives. Also, as the separation between the two pieces increases, λ increases and the wear rate decreases.

Using λ values and the separation between the pieces as a guide, there are three general categories of lubrication mechanisms. They are:

1. *Boundary lubrication* – This involves the range of lubricant film thicknesses where the peaks of the moving components are not totally separated from each other. It almost always occurs with slows speeds and frequently involves grease lubrication of parts slowly sliding over each other. In this area the additives are the most important component of the lubricant and the properties of the lubricated mating surfaces determine the friction and the wear rates. Applications include:
   a. Cast iron bearing surfaces, with no externally supplied lubrication, that rely on naturally occurring oxides and graphite

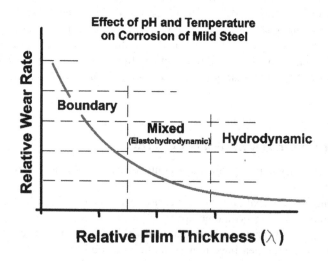

FIGURE 8.1   This shows the three basic lubrication regimes and how the wear rate decreases as the film thickness increases.

b. Slow moving plain bearings and heavily loaded gears with a significant proportion of extreme pressure (EP) or antiwear (AW) additives

c. Hydraulic cylinders

2. *Mixed film* (or elastohydrodynamic) lubrication is where most ball and roller bearings and high speed gears are found. Within this area keys to the wear rate are the viscosity, the cleanliness of the lubricant, and the deformation of the mating surfaces. With the extremely small film thickness, the additives that should be used in these lubricants are limited.

3. *Hydrodynamic* lubrication, also called full film lubrication is where the combination of lubricant and operating properties results in true separation of the pieces and very low wear rates result. It is in this range that minimum wear and maximum efficiency take place. Plain bearings are commonly found in large motors, internal combustion engines, and power generation equipment.

BALL AND ROLLER BEARING LUBRICATION (ELASTOHYDRODYNAMIC)

This is also called *mixed film* lubrication and occurs when the rolling elements of ball and roller bearings travel across a lubricated plane or high speed gear teeth mesh together. Figure 8.2 shows a side view of a typical rolling element bearing. At moderate to high speeds, at the incoming edge of the rolling contact, the lubricant inlet zone, the lubricant film is trapped and *viscosity conversion* takes place. At this point, the easy flowing liquid, the oil, is transformed to a semi- plastic fluid with a viscosity that can be a hundred or even a hundred thousand times higher than that same oil under normal atmospheric pressure.

Analyzing the Hertzian contact region of this rolling element contact, we see that the change in lubricant viscosity results in a separating film between the flattened portion of the element and the plane it is in contact with, greatly reducing the local stress and increasing the life of the bearing. Then as the lubricant leaves the contact region it instantly converts back to its unpressurized viscosity. The film thickness in this bearing is generally in the range of 0.0005 to 0.001 mm (0.00002 to 0.00004 in.)

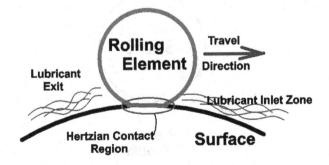

FIGURE 8.2 A view of a rolling element and its lubrication. What is amazing is that the oil viscosity will increase by 1000 or even a million-fold as it enters the Hertzian contact region, and then will go back to normal the instant it leaves. That change in viscosity allows the oil to physically deform and separate the rolling element from the surface.

(that is about 1/100 of a sheet of textbook paper) and the pressure is in the order of 2 GPa (≈300,000 psi). (The term *elastohydrodynamic* is used because the lubrication mechanism involves both elastic deformation of the rolling element and hydrodynamic action of the lubricant.)

HYDRODYNAMIC ACTION

The difference in internal lubrication between a rolling element bearing and a plain bearing is dramatic. The hydrodynamically lubricated plain bearing has much lower pressures, rarely exceeding 20 MPa (≈3000 psi), the clearances are in the order of 0.01 mm to 0.05 mm (≈0.0004 to 0.002″), and there is no viscosity conversion. (An exception to the 20 MPa [3000 psi] occurs in some railroad diesels that run as high as 45 MPa [6500 psi].) As shown in Figure 8.3, oil is fed into the bearing and trapped between the rotating shaft and the fixed outer bore. Primarily because of its viscosity, the trapped oil can't escape and it provides a low friction cushion for the shaft to operate on.

There is some frictional heating in a plain bearing and, as the oil grows hotter, the viscosity decreases and the leakage out the ends of the bearing increases. As long as the leakage rate doesn't exceed the feed rate, the bearing will continue to develop the necessary lubricating film.

BOUNDARY LUBRICATION

The third type of lubricated contact involves sliding contact between mating pieces and slow moving rolling element bearings. In the previous two examples the viscosity and cleanliness of the oil were critical. In this type of bearing, the lubricant

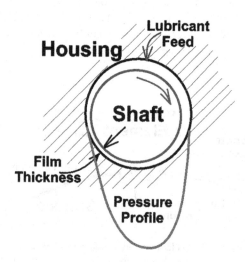

FIGURE 8.3 With hydrodynamic lubrication oil is fed into the bearing and as long as the leakage rate doesn't exceed the feed rate the two pieces are separated.

additives provide the operating surfaces that resist the wear and Figure 8.4 is a simplified view of what happens.

The action of the additives is critical to the effectiveness of these applications. Antiwear (AW) lubricants have a high proportion of polar fatty acid compounds one end of which is attracted to the metal's surface. These molecules line each side of the gap, helping to prevent contact between the opposing asperities. The effectiveness of these AW lubricants typically drops off dramatically as the local hot spot temperatures increase over 150°C (300°F). In those heavier, higher (local) temperature applications EP lubricants have far more load carrying ability and function in the two ways shown in Figure 8.5. Explaining those two methods:

1. Physical separation involves the use of additives such as molybdenum disulfide, graphite, and PTFE. The additives provide cushioning and physical separation between the two pieces to reduce the wear rate.
2. With chemical action, the additives react to the heat and pressure generated by contacts to develop coatings from the sulphur, chlorine, and phosphorous additive compounds that provide wear resistant surfaces.

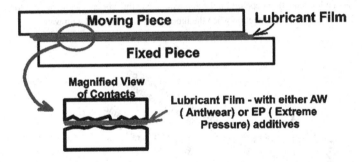

FIGURE 8.4 Sliding applications with boundary lubrication rely heavily on AW and EP additives to prevent wear by physically separating the two pieces.

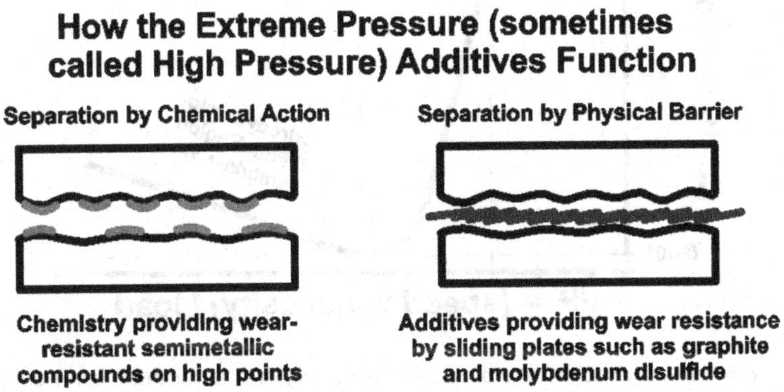

How the Extreme Pressure (sometimes called High Pressure) Additives Function

Separation by Chemical Action

Chemistry providing wear-resistant semimetallic compounds on high points

Separation by Physical Barrier

Additives providing wear resistance by sliding plates such as graphite and molybdenum disulfide

FIGURE 8.5 This shows the two common ways additives reduce wear in a boundary lubrication applications.

TABLE 8.1

Some Comparative Bearing Lubrication Data

Type of Contact	Typical Maximum Contact Pressures	Typical Film Thickness(mm/in)	Effect of Water
Rolling Element	2GPa/300,000 psi	0.001 mm/0.00004 in.	0.4% cuts life by 5 (Note 3)
Plain	20 MPa/3000 psi	0.025 mm/0.001in	Note 1
Sliding	1/2 metal's yield strength	0	Note 2

Notes:

1. From our discussions with technical lubrication personnel and equipment builders the primary problem caused by water contamination of plain bearing lubricants is that their corrosion resistance is reduced. The consensus of the folks we have talked with is that water contamination becomes harmful when there is more than 1.5% water in the oil.

2. The problem with sliding bearing lubrication and water involves the reaction with EP additives. AW additives are essentially immune to deterioration from water but EP additives such as sulphur and chlorine tend to form acids as they degrade. However, there are situations where, because both water and the AW additives are polar molecules, water can displace the AW molecules.

3. The actual life reduction depends on the application and the lubricant additives, but there will always be a substantial reduction when the lubricant is contaminated with water.

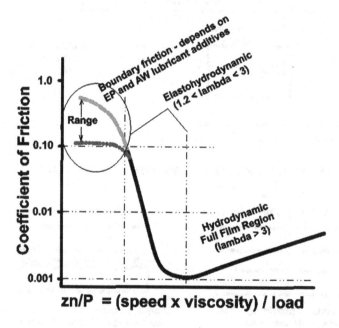

FIGURE 8.6 The Stribeck Curve describing the effect of film thickness on friction coefficient is based on the work of several scientists in the United States and Germany. (Adapted from Stribeck, *R. Die wesentlichen Eigenschaften der Gleit- und Rollenlager.* Zeit. des VDI 46, 1902.)

Most EP greases and oils combine a variety of additives to help perform their duties over a wide range of operating conditions.

Summarizing the bearing characteristics, Table 8.1 is a brief chart comparing the differences between three basic types of lubrication for contact surfaces.

THE STRIBECK CURVE

Reinforcing the concepts above, we can look at the Stribeck Curve in Figure 8.6, which describes how bearing friction is a function of the lubricant viscosity, the entrainment speed, and the load.

UNDERSTAND DIFFERENT LUBRICANTS

VISCOSITY MEASUREMENT AND VISCOSITY INDEX

The base oil is the "raw material" of the lubricant and viscosity is a measurement used to describe how readily that liquid flows. Critical is the lubricant's viscosity at the actual operating temperature.

There are several methods for measuring viscosity, and the most common one used in North America up until about the mid-1970s was the Saybolt Universal Second (SUS). This method has since been replaced by the metric centistokes, but the SUS is a good place to start because it is so easy to understand.

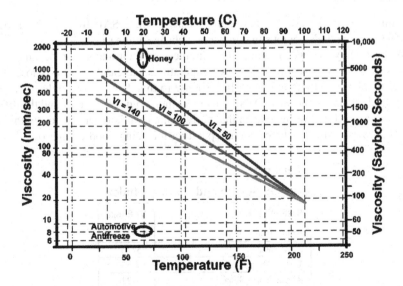

FIGURE 8.7   This shows how three oils with differing VIs change viscosity with temperature and compares them with both honey and antifreeze at room temperature.

Figure 8.7 is a chart with viscosity ranges as the vertical scales and the Saybolt scale on the right side. To check a lubricant's viscosity using a Saybolt viscometer:

1. A given quantity is put into a holding vessel where it is circulated and the temperature is held to within 0.1°F.
2. At the bottom of the holding vessel is a standard Saybolt orifice with a plug in it. Below the vessel and orifice is a stainless steel cup with an embossed line at the 60 ml mark.
3. When the orifice is opened and the oil begins to flow into the cup, the stop-watch is started.
4. When the liquid reaches the 60 ml mark, the watch is stopped.
5. If it takes 300 seconds to fill the cup, the viscosity is 300 Saybolt Seconds.

Looking at the figure, at a temperature of 20°C (68°F), the viscosity of automotive antifreeze is about 50 SUS and honey is somewhere between about 5000 and 10,000 SUS. (The system for measuring the kinematic viscosity of fluids using centistokes is similar and well proven but it doesn't have the simplicity of the SUS approach.) A fairly accurate conversion factor that can be used between about 30 and 2000 centi-stokes (cSt) is: cSt × 4.65 = SUS.

Another important concept to understand with liquid lubricants is the *viscosity index*. All liquids change viscosity as the temperature changes and some vary much more than others. The viscosity index (VI) is an artificial measurement that can be used to understand those differences and Figure 8.7 shows three oils that have the same viscosity at 99°C (210°F) but different VIs.

VISCOSITY INDEX

The three lines in Figure 8.7 show the viscosity of three SAE 10 (or ISO viscosity grade 32) oils and it's obvious that they don't change at the same rate.

Almost 100 years ago, as automobile engines and their lubrication became more sophisticated, it was realized that, as temperatures increase, some oils lose their viscosity more rapidly than others. In response to this the viscosity index (VI) was developed. Table 8.2 shows the VIs for five common crude oil sources and the differ-ence was a good part of the reason that early Pennsylvania based crudes, with their paraffinic base oils, developed their reputation for being good engine oils. Today

TABLE 8.2

Base Oil Source and Viscosity Index

Base Oil Source	VI Range
Pennsylvania (US)	90–110
North Sea	55–70
Texas (US)	25–55
Alaska North Slope	15–22
California (US)	5–15

most lubricating oils are still based on paraffinic crudes but almost all of the base oils have VIs of 95 or greater. In addition, there are many motor oils using synthesized base stocks that have relatively high VIs with some of the polyalphaolefins (PAOs) having VIs greater than 170.

Another view of how the different VIs result in different film thicknesses is shown in Figure 8.8, in which we've used three oils and set their viscosities equal at 20°C (68°F).

From this, we can see that oils with higher VIs have a great advantage in lubricating machinery that runs at elevated temperatures.

The classification of engine oil viscosities, i.e., SAE 10, 20, etc., is still universally measured by system used in SAE standard J 300a and the measurement of automotive gear oils, SAE 75W, 85W, 90 and 140, is usually measured by SAE standard J 306a. Most of the industrial oils we use follow the ISO classification outlined in ISO 3448 with 18 viscosity grades: 2, 3, 5, 7, 10, 15, 22, 32, 46, 68, 100, 150, 220, 320, 460, 680, 1000, and 1500. Except for the first four grades (2–7) the ISO Viscosity Grade number is the midpoint of the kinematic viscosity at 40°C (104°F) and the tolerance is +/– 10%, so the viscosity of an ISO 32 oil would be 32 cSt, +/– 10% or 28.8 cSt to 32.2 cSt.

Figure 8.9 puts much of what was covered above together in a chart that shows the approximate boundaries for both the SAE and ISO viscosity classification systems and how they vary with temperature.

These oils shown in this chart have a VI of 100, but there are many multiviscosity oils used primarily for internal combustion engines and hydraulic systems. In the case of multiviscosity motor oils, an SAE 10W-30 motor would act as an SAE 10W viscosity oil at −18°C (0°F) and an SAE 30 viscosity oil at 99°C (210°F).

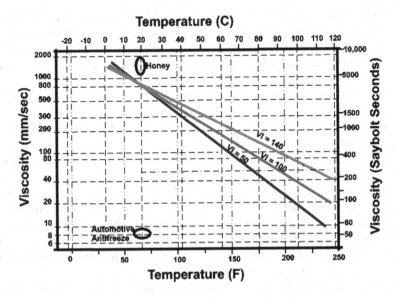

FIGURE 8.8 The three oils have the same viscosity at 20°C (68°F) but very different viscosities at elevated temperatures.

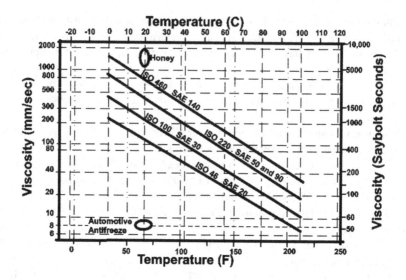

FIGURE 8.9 This graph is drawn for a 100 VI and shows a comparison of several ISO and SAE viscosity grades, but it doesn't show the low temperature effect of the SAE "W" grade lubricants.

BASE OILS

Recognizing the affect that the base oils have on finished engine oils, the American Petroleum Institute (API) has divided lubricant base stocks into five categories:

- Group I includes base stocks that contain less than 90% saturates; and/or more than 0.03% sulphur; and have a viscosity index at or above 80, but below 120.
- Group II includes base stocks that contain 90% saturates or greater; and/or only 0.03% sulphur or less; and have a viscosity index at or above 80, but below 120.
- Group III includes base stocks that contain 90% saturates or greater; and/or 0.03% sulphur or less; and with viscosity index of 120 or higher.
- Group IV includes PAOs, the most popular synthetic base stocks for formulating automotive engine oils.
- Group V includes all other base stocks not included in the first four groups.

Most base oils used for motor oils prior to 2001 were Group I. With the increased VI requirements of the newer automotive engine oils, most of the base oils are now Group III and IV. Gear lubricants are mostly Group II although there is increased penetration by Groups III and IV. Although Group I base oils are slowly going out of favor, their lower cost and improved additive solubility make them attractive for use in some applications such as marine engine lubricants.

Also, there are many specialized industrial applications, such as rotary screw air compressors, that specify Group V base oils, usually polyalkalene glycols (PAGs), because their superior elevated temperature and lubricating properties.

ADDITIVES

Regardless of the lubricant's viscosity, if it doesn't have the necessary chemical and physical properties, it won't be able do the job. The additives are what differentiates an automotive engine oil from a diesel engine oil or a steam engine oil, and they are critical to the lubricant's function.

An additive is a substance added to a base oil or a grease and most improve the performance characteristics of the lubricant but some, such as dyes and perfumes, are used to change the user's perception of the lubricant. Figure 8.10 shows the formulation of a motor oil and all finished oils are combinations of a base oils and several additives. Greases are a combination of a base oil, additives, and a thickener. (In the section where we talked about the composition of steels we mentioned that making a steel alloy was a little like making a cake where certain ingredients are added to get a specific product. A similar approach holds true for making a lubricant.)

Some of the more common additives are:

- *Oxidation inhibitors* – Reduce the formation of gums, sludges, and acids. Usually they are sulphur or nitrogen compounds but other chemicals like phenols and amines are also used. Oxidation inhibitors are found in most oils and greases except those with short expected lives. (If the oil doesn't have an oxidation inhibitor it slowly increases in viscosity and eventually becomes gummy. As an experiment, take some cooking oil and place it in a protected place and inspect it every couple of weeks. Eventually it'll become a sticky glue-like material.)
- *Viscosity index improvers* – Reduces the affect that temperature has on the fluidity of an oil. VI improvers reduce thinning at high temperatures and thickening at low temperatures and are found in many engine and multiviscosity hydraulic oils. Normally polymers such as polyisobutane are used. (In making a conventional engine oil with a Group 1 or Group II base oil, the base oil is usually a low viscosity, i.e., SAE 10 or so, then VI improvers are added so it behaves as though it were a 30 or 40 weight oil at high temperatures. The base oils used in synthetic and Group III motor oils have higher VIs than the Group II oils and don't require as much VI improver.)
- *Pour point depressants* – At low temperatures some of the paraffin molecules can form wax crystals if they aren't removed or chemically altered.

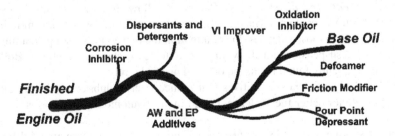

FIGURE 8.10   A little like a river on a physical map, a lubricant is the result of the base oil and the many possible additives.

Pour point depressants, usually aluminum soaps and aliphatic petroleum fractions, prevent the wax crystals from jelling, allowing the oil to flow more freely at low temperatures. Usually used in automotive, hydraulic oils, and some low viscosity gear oils that might see lower temperatures. (These are not needed with many synthetics.)

- *Defoamers* – Minimize the formation of foams by separating the bubbles from the oil. They are critical for engine oils, lighter gear oils, and hydraulic fluids, and are found in every type of oil subjected to mixing with air. The most common defoamers are liquid silicones (polydimethylsiloxanes), but too much defoamer will actually cause foam to form more readily.
- *Rust and corrosion inhibitors* – Reduce the probability of corrosion taking place in the lubricated areas. These vary from simple compounds to complex anionic polar film forming additives found in "paper machine oils" that prevent water from reaching the metal surfaces.
- *Emulsifiers and demulsifiers* – Depending on the specific additive they either promote or prevent the miscibility of water with the lubricant. *Demulsifers* are critical for many applications, such as turbines, where separation of water and the lubricant is important. In other applications, such as low speed wheel bearings that may be contaminated with water, *emulsifiers* are valuable in preventing corrosion and maintaining the lubricant action.
- *Antiwear additives* – These are generally polar fatty acid compounds that adhere to the opposing metallic surfaces, providing tenacious low friction separating layers between the surfaces. Above about 150°C (~300°F), local contact temperature they become ineffective.
- *Extreme pressure* – Outside North America these additives are known as high pressure additives, or HPs. EP oils are more effective in preventing wear than antiwear oils but are generally shorter lived. They are found in gear and hydraulic oils and there are two general types of EP additives:
    - Physical separators such a graphite, molybdenum disulfide, and Teflon that act to actually separate the opposing peaks. (There has been some discussion that these may interfere with the viscosity conversion that takes place in the load zone of rolling element bearings.)
    - Sulfur, chlorine and phosphorus that, when exposed to elevated temperatures from frictional contact, tend to form wear resistant compounds on the high spots.
- *Detergents and dispersants* – *Detergents* allow the oils to clean deposits from the contact surfaces. *Dispersants* prevent deposits from initially adhering to those surfaces. In combination the two additives keep contaminants in suspension and allow filtering systems to act more effectively. Both are frequently found in engine oils and hydraulic oils.
- *Friction modifiers* – Additives that reduce internal lubricant friction and provide for better fuel mileage. Found in most automotive engine oils.

In the latest issue of the AISE *Lubrication Engineers Manual* there are 64 different lubricant classifications (grades). The differences involve variations in both base oils and their additives and Table 8.3 shows the customary additive applications.

TABLE 8.3

Common Additives and Their Applications

Additive Category	Oxidation Inhibitor	VI Improver	Friction Modifier	Pour Point Depressant	Rust and Corrosion	Defoamer	Antiwear	Extreme Pressure	Detergent	Dispersant	Dye
Automotive engine	X	X	S	X	X	X	X		X	X	S
Automatic transmission fluid	X	X	X	X	X	X	X		X	X	X
Industrial gear oil	X	S		S	X	X	X	S			S
Industrial hydraulic fluid	X	S		S	X	X	X		X		S
Metalworking fluids	X				X	X	S	S	X		S
Turbine oils	X				X	X					X
Greases	X				X		X	X			X

X = Essentially always used in that lubricant S = Sometimes used in that lubricant

GREASES

Greases take that oil and add a thickener that serves as an oil reservoir. In hydro-dynamic and mixed film applications the thickener acts to slowly let the oil and additives wick out and provide the lubricant film, but in boundary lubrication applications the thickener joins with the oil and additives to provide the separating film.

The constituents of a grease are:

1. Additives – They typically range up to about 15% of the volume.
2. Thickeners – Ranges from a low of 4% to a maximum of about 20% of the volume.
3. The remaining fraction is the base oil.

There are several ways of categorizing greases with the most common North American system probably being the National Lubricating Grease Institute's (NLGI) numerical system for the grease stiffness shown in Table 8.4. This system refers to the pumpability or flowability of the grease but says nothing about its lubricating ability. For that we have to rely on data from the manufacturers and there is little in the way of publicly recognized standards that quantify the overall applicability of a grease for an application as a coupling grease or a ball bearing grease.

There are a tremendous number of standards used for testing of greases such as ASTM D-1742 "Oil Separation from Lubricating Grease During Storage", that apply to one specific part of a grease's function or storage. The only specific grease performance standard that I know of is the NLGI GC-LB, and that's a very general specification for automotive chassis and wheel bearing grease.

TABLE 8.4
National Lubricating Grease Institute (NLGI) Grease Grades

NLGI Grades	Dart Penetration (0.01 mm)	Consistency and Typical Applications
000	445–475	These are very soft greases and tend to flow. Similar to heavy oils, they are generally used to lubricate large slow gears.
00	400–430	
0	355–385	
1	310–340	These are usually used for lubricating rolling element bearings, most likely with #2 grease in motors and pillow blocks. At higher temperatures, there are times that #3 greases are used.
2	265–295	
3	220–250	
4	175–205	"Block" greases, very hard brick like greases usually used for sliding or low speed rotating applications such as rotary vessel tire and wheel surfaces, and so on.
5	130–160	
6	85–115	

When a grease is specified for an application some points to keep in mind are that:

1. For ball and roller bearings:
   a. The viscosity of the base oil is critical to providing the supporting film in the Hertzian contact region.
   b. Most of the data we have seen has shown that the addition of EP additives generally reduces the life of the bearings by about 30%.
2. For gear and grid couplings, above about 300 rpm it is a necessity that a specified coupling grease be used for long life and reduced relubrication intervals. (Above that speed, the centrifugal force is enough to separate the oil from the thickener; then the heavier thickener slowly moves outward, displacing the oil and additives in the areas that need lubrication.)
3. For gears, the grease should be very fluid and contain a significant amount of AW or EP additives.

LUBRICANT APPLICATIONS

ROLLING ELEMENT BEARINGS

Early in this chapter there was a section explaining how lubricants function in the contact area between the rolling element and the race. In this section we'll expand on that and add the action of the lubricant in the rest of the bearing.

Figure 8.11 shows a cross-section view of a rolling element bearing. In this example the bearing is under a pure radial load and as such:

1. When the rolling elements are in the load zone they are trapped between the two rings and they drive the cage.
2. As soon as the rolling elements leave the load zone the cage *drives them.*

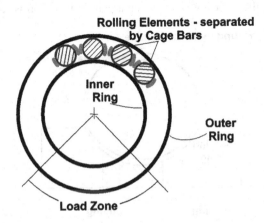

FIGURE 8.11 A cross-section view of a rolling element (ball or roller) bearing with a vertical load on it.

This accounts for one of the lubricant requirements and that is the need to provide cushioning between the elements and the cage as they go in and out of the load zone.

A second lubricant requirement comes into play when the rolling element returns to the load zone. There is always some minute misalignment between the two rings and there is always some friction between the rolling elements and the cage in the unloaded portion of the bearing. As a result, there is an almost instantaneous acceleration of the element as it returns to the load zone and, if the lubricant film isn't adequate, skidding occurs. (This is similar to what happens when an airplane tire first contacts the runway on landing.) With time this skidding roughens the rolling path and premature failure results.

(The description above only applies to bearings that have no or very little thrust preload. One of the advantages of a thrust loaded bearing is that the rollers do not experience this acceleration and deceleration and the resultant skidding. One sure way to tell if these actions have affected the life of a bearing is to look at the load zones after removing the bearing from operation.)

The last need for lubrication is shown in Figure 8.12, where a heavily loaded and elastically deformed ball is shown in contact with a race that was perfectly machined to contact a lightly loaded ball. The no-longer-round ball has several areas of rolling and skidding and a separating lubrication film is required to prevent damage.

The Pressure Viscosity Coefficient

Much of what we see in ball and roller bearing lubrication also applies to gear lubrication, but with the gears there is also a substantial amount of sliding action. Many lightly loaded bearings and gears could probably be well lubricated by clean corn oil, but for heavily loaded or elevated temperature conditions, an understanding the pressure viscosity coefficient is very helpful. (However, one of the problems with the idea of using corn oil is that it tends to oxidize and eventually becomes sticky and gummy, and this is one of the major areas of biolubricant research.)

As mentioned in conjunction with the Hertzian contact area, when fluids are exposed to high pressures their viscosity increases. Figure 8.13 is a chart showing a summary of how the viscosity varies when a plain mineral oil is subjected to these

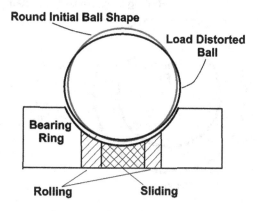

FIGURE 8.12 This shows the distortion that occurs as both the load and the elastohydrodynamic forces deform the ball and skidding occurs.

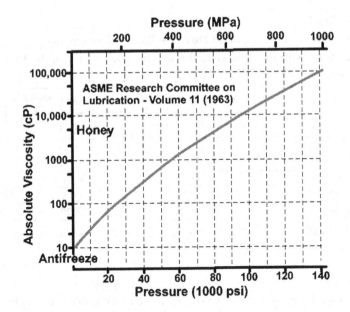

FIGURE 8.13 This shows how the viscosity of a light oil changes when subjected to pressure. This chart ends at about 1 GPa (~145 ksi), while pressures in ball and roller bearings commonly exceed 2 GPa (~390 ksi).

pressures. We included the honey and antifreeze examples from the earlier chart and, when you consider that a heavily loaded rolling element bearing may see pressures over 2 GPA (≈300,000 psi), the change in viscosity is tremendous! At room temperature, this oil goes from being slightly more viscous than antifreeze to the point where, at less than half the peak pressures seen in some bearings, it is already more than 100 times more viscous than honey. The research shown in this graph was conducted over 60 years ago and since that time material science advances have enabled even higher pressure analysis that shows the viscosity increases by another10^5! As that easy flowing oil enters the rolling element contact area, it instantaneously converts to a glass-like solid, and then just as rapidly, reverts to its original viscosity on exiting.

The pressure-viscosity (PV) coefficient is a measure of the relative change in viscosity of lubricants as the pressure increase and is largely a function of the base oil used in the lubricant. The importance of this is that some base oils have substantially greater PVs and as a result have thicker viscosity conversion films at the operating temperatures and pressures. Table 8.5 gives a selection of lubricating oils and their relative lubricant film thickness at both 25°C (76°F) and 80°C (176°F) @ 2000 bar (≈29,000 psi). But there have to be two cautions about using this table:

1. The range of pressures in the chart does not go very high, i.e., 2000 bar when bearings and gears sometimes operate at seven to ten times that pressure.
2. The PV coefficient refers to the increase in viscosity of that lubricant at that temperature and pressure. If the application has an oil with a low VI and it is running at the same conditions as the high VI oil, the latter starts with a big advantage.

TABLE 8.5

Showing the Variation in How the Base Oil Viscosities Change with Pressure (i.e., the Relative Difference in the PV Coefficients)

Oil Type	Viscosity Range @ 2000 bar/Atmospheric Viscosity at 25°C (76°F)	Viscosity Range @ 2000 bar/Atmospheric Viscosity at 80°C (176°F)
Paraffinic mineral	15–100	10–30
Napthenic mineral	150–800	40–70
PAO's	10–50	8–20
Polyether oils	9–30	7–13
Ester oils	20–50	12–20

Data based on information from *Lubricants and Related Products,* by Dieter Klamann, published by Verlag Chemie, Weinheim, FRG, 1984.

In addition to that shown in the chart, there are other features that help account for the improved performance of PAOs and PAGs (polyalkylene glycols) in heavily loaded applications above 75°C (168°F). There is data that shows their PV coefficients tend to increase at the higher temperatures and they also have better thermal conductivity than conventional mineral oils. (The effect of the improved thermal conductivity is these lubricants tend to run cooler with the result that they have a higher viscosity and thicker lubricant films.)

Lubricant Films and the Effects of Contamination

There have been many studies that have shown that water contamination reduces the fatigue life of bearings. From 1960 to date, we are aware of at least seven extensive analyses, none of which had the operating conditions duplicating other studies. Based on them, we feel that there can be a consistent reduction in life at water concentrations as low as 0.01% (100 ppm) and recommend that water levels be kept below half of that (50 ppm).

The problems that water presents are multiple, and:

- As a polar molecule, when it is put under pressure unlike lubricant oils it does not increase in viscosity. When it contaminates the typical hydrocarbon lubricant and is put under pressure it tends to vaporize. This phase change, a tremendous change in volume, removes some of the hydrocarbon lubricant and damages the cushioning layer in the Hertzian contact zone.
- When the bearing is shut down the water leads to corrosion.
- As a polar molecule, it tends to displace the polar antiwear additives on the metal surfaces.

Every text on bearings has charts like that in Figure 8.14 showing that as the film thickness increases bearing life increases. From a practical standpoint, probably the

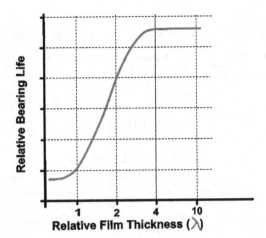

FIGURE 8.14 Relative film thickness compared with relative bearing life.

most important thing that can be done is to pay attention to the lubricant temperatures and the minimum viscosity recommendations below.

Developing a Relubrication Program

To determine the regreasing frequency, use Table 8.6 and Figure 8.15. They will give a starting point for planning the relubrication of your ball and roller bearings. However, realize that this procedure is based on the needs of the bearings and many times relubrication is used as a supplementary seal.

1. Start with Figure 8.15. Knowing the bearing rpm and the bearing bore size, draw a horizontal line from the rpm to the bore, then draw a vertical line to determine the idealized relubrication frequency.
2. Next, using Table 8.6, determine the factor for each of the operating conditions.
3. Calculate the suggested actual relubrication frequency by multiplying the result from Figure 8.15 by each of the factors. Therefore F_{act}=Figure 8.15 value $\times F_t \times F_c \times F_m \times F_y \times F_p \times F_d$

As an example, if there was a fan with a horizontal 75 mm (3″) shaft outdoors on the roof of a building running at 1770 rpm the calculation would be

- From Figure 8.15 – 75 mm (3″) shaft @ 1770 rpm=8000 hours
- F_t=1.0 – Housing temperature was below 65°C (150°F)
- F_c=0.4 – Light abrasive dust on the building roof
- F_m=1.0 – Humidity usually below 70%
- F_y=0.6 – Vibration range is 5 to 10 mm/sec (0.2 to 0.4 ips) peak vel
- F_p=1.0 – Horizontal shaft
- F_d=0.2 – The shaft has spherical roller bearings

TABLE 8.6

Factors Used for Calculating Regreasing Intervals

Condition	Typical Operating Condition	Factor
Housing temperature F_t	Below 65°C (150°F)	1.0
	65 to 80°C (150 to 175°F)	0.5
	80 to 95°C (175 to 200°F)	0.2
	Above 95°C (200°F)	0.1
Contamination F_c	Light nonabrasive dust	1.0
	Heavy nonabrasive dust	0.7
	Light abrasive dust	0.4
	Heavy abrasive dust	0.2
Moisture F_m	Humidity almost always below 70%	1.0
	Humidity between 70% and 90%	0.7
	Occasional condensation	0.4
	Occasional water on housing	0.1
Vibration F_v	Less than 5 mm/sec (2 ips) peak velocity	1.0
	5 to 10 mm/sec (0.2 to 0.4 ips) peak vel.	0.6
	Above 10 mm/sec (0.4 ips) peak	0.3
Position F_p	Horizontal bore centerline	1.0
	Bore centerline at 45° to horizon	0.5
	Vertical centerline	0.3
Bearing type F_d	Ball radial and thrust bearings	1.0
	Cylindrical and needle roller bearings	0.5
	Tapered and spherical roller bearings	0.2
	Roller thrust bearings	0.05

The recommended regreasing frequency would then be:

$$F_{act} = 8000 \times 1.0 \times 0.4 \times 1.0 \times 0.6 \times 1.0 \times 0.2 = 384 \text{ operating hours}$$

If the fan was running 24 hours per day, seven days per week, that would call for regreasing every two weeks.

The suggested relubrication quantity would depend on the size of the bearing but a good starting point can be calculated by:

Metric dimentions – grease fluid volume (ml) = 0.005 × bearing width (mm) × bearing OD (mm)

or *– grease fluid volume (ml) = ½ × bearing width (cm) × bearing OD (cm)*

American dimensions – grease fluid volume (ounces) = 0.1 × bearing width (inches) × bearing OD (inches)

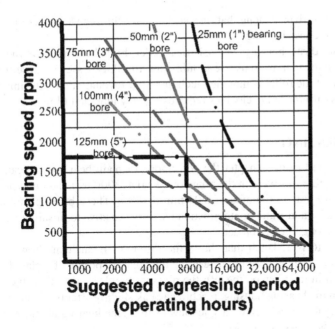

**Suggested regreasing period
(operating hours)**

FIGURE 8.15 This is the first step in developing a regreasing plan for ball and roller bearings.

Table 8.7 shows the minimum suggested viscosities for several common bearings and gears. One important point about this chart is that at low gear and bearing speeds there isn't enough velocity to cause viscosity conversion and these applications have to be treated as though they involved sliding contact. Some other general notes are:

1. Roller bearings such as cylindrical, spherical, and tapered rollers require a greater film thickness than ball bearings because they are more sensitive to surface variations, i.e., instead of a point contact, they have line contact and

**TABLE 8.7
Suggested Minimum Viscosities at Operating Temperatures for
Typical Industrial Machinery**

Minimum Viscosity (mm/sec)	Equipment
13	Ball bearings, hydraulic systems
23	Spherical, tapered, cylindrical roller bearings, lightly loaded gears
35	Roller thrust bearings
40	General spur and helical gears
70	Worm reducers

any disturbance along that line, such as a machining variation or a speck of dirt, reduces bearing life.

2. Roller thrust bearings and spherical roller thrust bearings have radial contact lines that are sensitive to lubricant film thicknesses because the outer edge of the roller is moving much more per revolution than the inner edge. This difference in velocities requires a heavy lubricant film.

PLAIN BEARINGS

In the first part of this chapter the description of plain bearings indicated that the maximum pressures were in the range of 20 MPa (3000 psi) and the typical operating clearance was in the area of 0.03 mm (0.001″). These bearings are tolerant of small particles and water contamination but very sensitive to misalignment because they have a long line of contact.

Figure 8.16 is a sketch that shows the results of Beauchamp Tower's series of landmark experiments used to determine the pressure profile of a plain bearing. As can be seen in the view on the left, the pressure builds as a result of the rotation of the shaft in the bearing housing, reaching a peak slightly after the center of the bearing. The right diagram shows a side view of the same bearing along with the three profiles of the pressures across it.

From these drawings it is easy to see the disastrous effects of not only misalignment but also of putting an oil groove, or other feature that allows lubricant escape from the loaded field.

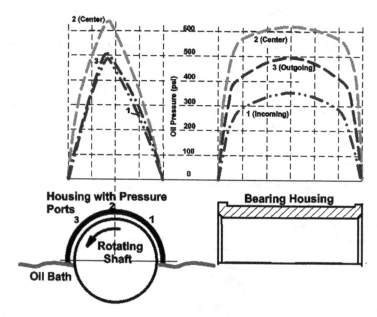

FIGURE 8.16 A sketch of Beauchamp Tower's original experiment where he determined how plain bearing lubrication actually works. The bearing was 4″ diameter × 6″ long (100 mm OD and 150 mm long) and ran in a light oil bath.

Earlier in the chapter we wrote, "There is some frictional heating in a plain bearing and, as the oil grows hotter the viscosity decreases and the leakage out the ends of the bearing increases". When a plain bearing is more heavily loaded, it will develop more frictional heat. That will decrease the oil viscosity and increase the leakage rate, but as long as there is adequate feed to the bearing, the bearing will support the load. (Calculating the heat generation and resultant viscosity is very complicated and beyond the scope of this text.)

LUBRICATION SUMMARY

Lubrication appears fairly simple but the details are extremely complicated. In this chapter we have tried to explain some of the more important points of fundamental lubrication mechanisms with emphasis on ball and roller bearings. Unfortunately, selecting the ideal lubricant for a plain bearing depends on the operating temperature, the surface speed, bearing loads, and the bearing dimensions and the complexity is beyond capability of this text.

DEFINING WEAR

Wear is the undesired removal of material by contact with another material. There are three different basic wear mechanisms and from what we have seen they typically account for between 3 and 7% of the total maintenance cost in the average plant. However, in some plants, like difficult mining operations, they may amount to almost half of the total cost.

There is disagreement within the failure analysis community as to the wear categories. Some folks say that there are five different mechanisms, surface fatigue, fretting, adhesion, abrasion, and erosion. Others include fretting in with corrosion and lump surface fatigue with the other fatigue mechanisms. We have chosen to include both surface fatigue and fretting in our general discussion of fatigue with additional details on surface fatigue in the section on bearings and additional information on fretting in both the section on bearings and the section on corrosion. So, this section will address adhesion, abrasion, and erosion.

From a practical standpoint, one of the difficulties with wear is that it is almost impossible to diagnose the specifics of the problem without a metallurgical examination. It isn't hard to tell that something has worn away but identification of precisely whether it is adhesion, abrasion, or erosion needs that microscopic examination. Two other situations that add to the complexity of the problem are:

1. Frequently a situation will involve synergistic mechanisms, i.e., more than one wear mechanism is responsible for the loss of material.
2. It is not uncommon that wear and corrosion mechanisms coexist and act together to accelerate equipment deterioration.

More than once we've seen people working and working to try to solve a wear problem that turns out to actually be a corrosion problem.

Adhesive wear occurs by transferring material from one surface to another by solid-phase welding. ("Solid phase" means the pieces are welded together at what is essentially room temperature and not at the elevated melting point of the metal.) Some common terms used to describe adhesive wear are scoring, galling, scuffing, and seizing.

When a stainless steel bolt seizes during assembly or a gear tooth such as that shown in Photo 8.1 has radial drag lines, adhesive wear has been there and the pieces have welded themselves together at close to ambient temperatures.

In the dedendum of the tooth shown in the photo the opposing tooth contact has been severe enough that small areas have been welded and then torn out.

Photo 8.2 is of the mating surface of a 30 mm (~1.25″) stainless steel nut and the contact face has been badly damaged by adhesive wear. Most people would refer to this as galling and it is a very common problem with unlubricated austenitic (300 series) stainless steel fasteners.

Detection – The heat of welding almost always causes changes to the material's microstructure that can be seen by a metallurgical examination.

Prevention usually involves using materials that are insoluble in each other, i.e., cannot weld to each other, such as cast iron and hardened steel, or the use of special surface coatings or EP lubricants. Other alternatives involve reducing the local temperature to reduce the possibility of welding and using materials with very smooth surfaces. (This latter approach may sound impractical to some, but it is actually one of the methods used to increase the power throughput in precision gear sets.)

Common practice is, when gears or similar pieces of equipment are meshing against each other and the materials are similar (e.g., they are both steel), the hardness of the pieces should be separated by at least 10 Rockwell points.

PHOTO 8.1 A close up of a gear tooth with severe adhesive wear. The mating tooth has rubbed against this one and torn pieces out of it through solid-phase welding.

PHOTO 8.2   The heavily galled mating surface of a 300 series stainless steel nut.

Photo 8.3 shows a section of a broken bolt that has welded itself to the flange and it initially looks like an exception to this rule. In this example the bolt and the flange were separated by 25 HRC points but three things happened to allow them to weld together:

1. The pieces were at about 200°C (≈395°F), so they started at temperatures well above ambient.
2. They were not in intermittent contact like the gear teeth. (The advantage of the intermittent contact is that it allows the tooth surfaces to cool before another contact is made.)
3. They had very little (or no) lubricant on them that could prevent the welding.

(With the assembly of austenitic stainless fasteners, or stainless devices that could have rubbing contact with other austenitic pieces, it is critical that either the surfaces are lubricated/coated or that galling-resistant specialty stainless steels, such as the Nitronics, are used.)

PHOTO 8.3   This ~16 mm (5/8″) bolt has "seized" and welded itself to the flange after being tightened by hand, using a short-handled ratchet wrench. We had to saw cut the flange to open it up to show exactly what had happened and the missing next-to-last thread is obvious.

Abrasive wear involves three categories, gouging, high stress grinding, and low stress scratching. that differ slightly in their mechanisms, but all involve material removal from a solid surface by particles sliding along the surface. Some other terms used to describe abrasive wear include scouring and erosion.

Gouging results when large particles are removed from the surface leaving grooves or pits. A good example is the teeth of a power shovel in a quarry handling large heavy rocks. This requires materials that are relatively hard and also have good fracture toughness such as "Hadfield Steel" or tempered medium carbon alloy steels. To the best of our knowledge there is no accurate laboratory test correlation.

High stress grinding (HSG) occurs when there is either plastic flow of ductile or fracture of brittle target materials when the load is high enough to cause fracture of the abrasive medium. The worn surfaces show scratch marks where the material was removed. This is typical of industrial grinding of minerals in ball and rod mills, with fine particles under heavy pressure. Impact resistance is not needed, but the target materials need some fracture toughness and hardened white irons and high carbon steels can be economically used in most applications. As a laboratory test, the wet sand erosion test seems to be the most applicable.

Low stress scratching (LSG) happens when there is no impact loading and the unit pressures are relatively low. It is what would be seen with a plow in sandy soils where the repeated contact with soil particles would slowly remove target material. In these conditions the matrix is not critical, as toughness is not important, and materials with hard carbides are generally used. As a laboratory test, the dry sand erosion test is a good source of comparative data.

Detection of abrasive wear in metals often requires the use of metallography to look at the outermost layer of the grain structure to see if it is deformed from the severe contact stress. Other times, such as in Photo 8.4 it is relatively obvious that the piece has been abraded away. This photo shows a tremendously worn cast iron chain sprocket from a wood yard and the source of the wood was a sandy area.

PHOTO 8.4   This wood yard sprocket has been heavily worn by the abrasive sand carried by the logs.

PHOTO 8.5 A great example of using hardfaced weld alloys to improve the wear resistance of this drag line bucket. Note that different alloys may be used depending on the material being mined. (A different overlay would be used for sand than would be used in hard rock.)

In nonmetallic materials abrasive wear usually shows grooves or gouges, frequently with small scrapings (or shards) of parent material alongside the damage path (see Photo 8.6).

Abrasion prevention can involve several approaches;

1. In LSG the hardness of the target material is not as important as the structure. For example, a hardened steel and a hardened cast iron may have identical HRC 58 hardness readings but the cast iron with its extremely hard carbide "islands" will far outlast the hardened steel in many applications. (In the cast iron the carbides are much harder than the HRC 58 overall hardness reading and they act to shield the surrounding matrix from the action of the wear particles. The steel is essentially homogenous and doesn't have the same ability.)

2. In both gouging and HSG harder target materials with improved fracture toughness are frequently used. Hardfacing, welding an extremely hard and abrasion resistant material on the surface of a target area, is frequently used in the mining and materials processing industries to provide local protection as shown in Photo 8.5. This is a section of a dragline bucket and the weld overlays are laid on top of the bucket steel in a pattern to reduce the abrasive wear. These overlays are typically very brittle but are supported by the tougher and more ductile parent material.

3. Reduce the pressure and velocity of the abrading particles.

Textbooks frequently mention that changing the abrading particle shape can improve a wear situation but that is usually impossible to do in an operating plant. However, one good lesson on the importance of particle shape in an abrasive or erosive application occurred when we replaced some steel pipes with HDPE piping. These pipes carried sand from our crusher to the settling pond and the steel pipes repeatedly wore out after only two years. Another division of our company made the resins for HDPE pipe and their data showed that HDPE would outlast steel about 5 to 1 in LSG from sand in water, so we installed several sections of 12″ HDPE. About two months after the installation we decided to use ultrasonic thickness testing to check on the wear rate of the pipe and as soon as we looked at the thickness data, we immediately told them to change the pipe. The sand particles were the same size but the wear tests had been conducted with river sand which is old and smooth compared to the particles that had just been fractured in the crusher and those sharp edges had attacked the new pipe with a vengeance. When they removed the pipe after operating for only three months, there were areas that were down to less than 10% of the original thickness. If anybody asks you, our "research" shows steel outlasts HDPE by about 6 to 1 when used on new sand!

Erosion generally involves flowing loose particles attacking a surface but it is sometimes identified by local terms. The three general categories are:

1. *Airborne erosion* from solid particles carried by the gas flowstream. The particles usually range in size from 0.0002″ to 0.02″ and the typical airstream's velocity is over 15 fps.
2. *Liquid erosion* from solid particles carried by the liquid flowstream. The solid particles are usually over 0.002″ diameter with flowstream velocities over 6 fps.
3. *Liquid erosion* from liquid droplets over 0.010″ carried by a gas stream with velocities dependent on flowstream components.

Erosion is usually identified by a combination of the distinctive turbulent flow patterns left on the attacked material and the knowledge of the impinging stream.

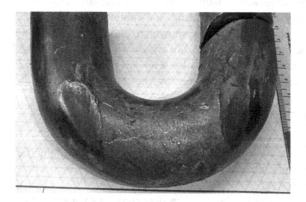

PHOTO 8.6 A ~38 mm (1.5″) thick chain link from a mining operation with two badly abraded spots where the chain didn't have enough clearance. The pushed up material is proof of the gouging abrasion, and if the chain had been harder it wouldn't have worn so much. (But that might cause other problems.)

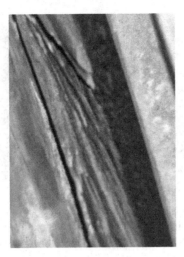

PHOTO 8.7   This shows the worn steel diaphragm in a centrifugal compressor. The wear was from the high velocity water particles in the gas stream.

PHOTO 8.8   A closeup of the eroded worn surface of a pump impeller. (The flow was from left to right.)

Photo 8.7 shows a diaphragm out of a large centrifugal compressor with substantial erosion damage from a wet flowstream at 5000 rpm on a 450 mm radius. Photo 8.8 shows the surface of a pump impeller that suffered erosion from solid particles in the flowstream.

Photo 8.9 is interesting because it looks as though it could be solely a corrosion problem, but actually a major contributor was erosion. This is a copper hose fitting that was used to carry some slightly acidic water from a pumping operation in a national forest. The leak started out as an occasional drip and within a couple of weeks was a steady spray. The path across the fitting is easy to see, while the oxide in the unattacked areas speaks to the corrosion resistance of the copper.

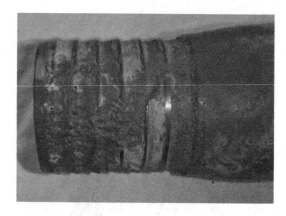

PHOTO 8.9   A copper fitting with some corrosion, but mainly erosion damage. It should be noted that, if there had been no corrosion, the leak would never have happened.

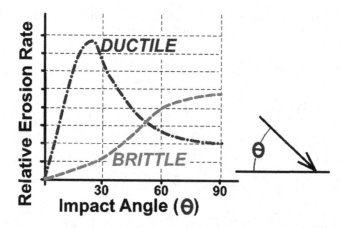

FIGURE 8.17   This shows how the impingement angle affects the wear rotes of different materials and is based on the work of J.G.A. Bitter.

AIRBORNE EROSION

Two of the interesting properties of airborne erosion are that not only do the target and impinging stream angles have an immense effect on the relative wear rates as shown in Figure 8.17, but also the particle velocities have an equally impressive effect. From this we can look at the relative flow rates and predict almost exactly what effect a change in process flow will have on erosive wear rates.

Another interesting curve is shown in Figure 8.18. That shows some of the work of Wellinger and Uetz who found that the wear rate of a target depends on the hardness of the abrading particles. Looking at it's interesting to see that even particles as soft as chalk can eventually wear away hardened bearing steel.

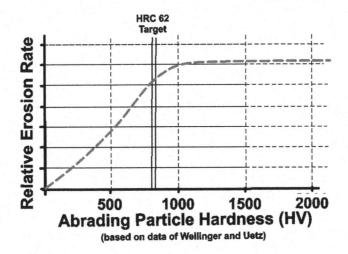

FIGURE 8.18 The relative wear rate of a target depends on the hardness of the abrading airborne particles up to a point, but beyond that it is relatively consistent.

SUMMARY COMMENTS ON WEAR FAILURES

It is usually difficult to understand exactly what is happening with a wear failure without a metallographic examination. We've seen numerous examples of changes in materials that haven't solved the problem and sometimes made the situation worse. However, with airborne and liquid erosion, changes in operating conditions can have very predictable effects on component life.

BIBLIOGRAPHY

Fuller, Dudley D., Theory *and Practice of Lubrication for Engineers* (2nd edition), John Wiley & Sons, Inc., 1984, ISBN: 0-471-04703-1.

Klamann, Dieter, *Lubricants and Related Products*, Verlag Chemie, Deerfield Beach, FL, 1984, ISBN: 0-85973-177-0.

The Lubrication Engineer's Manual, American Iron and Steel Engineers, Pittsburgh, PA, 1996, ISBN: 0-930767-01-2.

Tribology and Lubrication Technology, published by the Society of Tribologists and Lubrication Engineers, Park Ridge, IL (numerous technical articles).

9 Belt Drives

Belts were developed to provide a means for transmitting power from the prime mover – a water wheel and later a steam engine – to the point where it was actually needed. The first belts were used in water-powered mills and they were generally flat, made from leather, and ran on wooden pulleys that had a very slight crown to keep the belt in position. In those early days, regardless of the power source, a mill would have a "line shaft" driven by the prime mover. Then, off the main line shaft, there would be many flat belts, each supplying a smaller line shaft, a machine, or a part of the process. (One of the truly impressive sights of my early days in industry was watching the "head operator" using a special stick-like tool to put a 24″ wide, 30-foot long flat belt on a line shaft pulley rotating at 500 rpm. A huge motor drove the main line shaft, which in turn drove 15 other machines.)

Before the modern variable speed motors were introduced, AC motors only ran at given fractions of their synchronous speed (3600, 1800, 1200, 900 rpm, etc., in North America, and 3000, 1500, 1000, 750 rpm, etc., in Europe and those areas with 50 cycle power). If the machine needed an operating speed other than one of these, there had to be some method of changing the speed. This could be done with gears or chains but, compared to V-belts, they are expensive, noisy, complicated, and require lubrication. (However, both offer precise speed regulation, something that is impossible with a V-belt or a flat belt.)

V-belts were first used commercially shortly before 1920. In the United States, Gates made the first industrial V-belts while Dayco (now Carlisle) pioneered the automotive V-belt, and the first synchronous belts were developed in the 1940s to assure proper timing of the needles and bobbins in industrial sewing machines and are now found in many applications In the last 60 or so years there have been tremendous changes in belt technology with the "modern V-belts" showing up in the 1960s, raw-edged notched V-belts and the first serpentine belts showed up in the 1970s, and many variations on the original square-toothed synchronous belts.

BELT DESIGN

Figure 9.1 is a sketch showing the major V-belt components. The original construction used tension members made from cotton or linen, the elastomer was natural rubber, and the canvas wrapping was really necessary to keep the parts from separating. Through the 1950s and 1960s, the tension members were generally made from rayon. Today most tension members are made from polyester though many alternatives are available for special purposes. The elastomer is a synthetic rubber compound, usually either SBR or neoprene, and wrappers are becoming less common and generally used in specific applications.

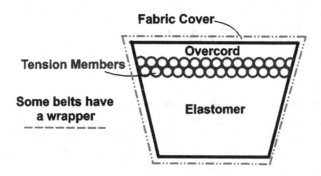

FIGURE 9.1   A cross-section view of the typical V-belt.

The classical V-belts were developed years ago using the technology of that era and were a great improvement over flat belts, not only because they were self-aligning, but also because the way they function reduces the amount of load on the supporting shaft bearings. However, one of the weaknesses of the original V-belt design was that the center tension member cords didn't carry a significant portion of the load while their flexure created heat. To improve this, the "modern" 3V, 5V, and 8V belts were introduced in the 1960s, and they have narrower and deeper profiles than the classical belts. Figure 9.2 shows how the different North American belts compare in their relative sizes and load capacity.

There are also a series of metric V-belts that follow the British Standard BS3790 (see Table 9.1) and are similar to the modern V-belts but with slightly different dimensions.

Another change in the construction of North American V-belts involved the introduction of "molded notch" belts that Dayco introduced in the 1970s. These have the same basic dimensions and act in the same manner as modern and classical V-belts except some of the material in bottom of the belt has been notched out, allowing the belts to wrap more easily around sheaves. Although it is true that these belts can be used on smaller diameter sheaves, they also run cooler on standard size sheaves and take less energy to flex. (There is a group of power transmission [PT] belts used on variable speed machinery, such as snowmobiles and some ATV's, that looks a little like a molded notch belt. However, putting them side-by-side it can be seen that the PT belt is much wider, thicker, and more rigid than a V-belt.)

In addition to the obvious difference between raw-edged and wrapped V-belts, other internal differences are:

- Most wrapped belts use SBR as their elastomer while raw-edged belts use neoprene, which has better heat resistance and stability.
- For a given width, raw-edged belts have more tension member cords, are generally stronger, more temperature resistant, and more efficient – but they cost about 25% more.
- With the different materials, dimensions, and friction coefficients, raw-edged belts are much more sensitive to badly worn sheaves and will slip more readily.

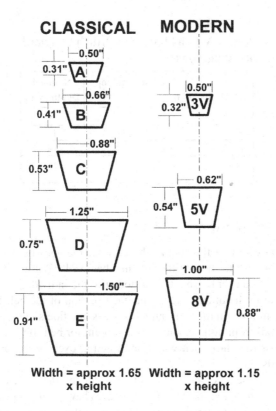

CLASSICAL   MODERN

Width = approx 1.65   Width = approx 1.15
      x height   x height

FIGURE 9.2   Showing the difference between the original and "modern" V-belts used in North America.

Photo 9.1 shows a pair of "banded V-belt" sets and the differences in basic construction and size are obvious. The set on the left is a pair of conventional wrapped B belts while the set on the right is a pair of raw-edged 3V belts. Although there is a huge difference in size, the two sets have almost the same capacity.

There are many other belt shapes ... double sided V, V-ribbed (serpentine), etc ... all of which have applications and all of which can be analyzed with many of the same approaches used on V-belts.

Some of the internal differences in synchronous belts are:

- The original synchronous belts, those with square or rectangular teeth, have fiberglass cord tension members while the higher performance types, with their rounded teeth, use cords made from aramid fibers such as Kevlar™ and Spectra™.
- The aramid fiber belts require more initial tension than fiberglass belts because there is some initial aramid fiber stretch, but it is far less than what would be seen with a polyester member. Also, polyester tension members continually stretch to some degree while the later stretch of the aramid fibers is almost imperceptible.
- Both types of synchronous belts use neoprene as the elastomer.

TABLE 9.1

Metric V-Belts Based on British Standard BS 3790-1995

Belt Designation	Nominal Top Width (mm)	Nominal Depth (mm)
SPZ	10	8
SPA	13	10
SPB	17	14
SPC	22	18

BELT OPERATION OVERVIEW

As a V-belt wraps around the sheave the elastomer bulges outward, transmitting power from the sheave to the belt tension member cords. Within the belt, the loads should be transmitted evenly from the elastomer to the individual cords, which then transfers the load to the other sheave, and each section of the belt is stressed hundreds of times a minute. In this continuous process, the fibers in the tension member will eventually fail from fatigue, and a critical point for belt life is even distribution of the load across the entire belt width, eliminating excessive stress in any of the individual strands.

(Analyzing the sheaves, the angles on the side of the V-belt groove change depending on the size of the sheave. Larger sheaves have straighter sides because the belt is wrapped over a larger radius and the sides of the belts don't deform outward as much.)

Flat belt operation initially appears similar to V-belt operation but, because the flat belt doesn't have the wedging action of the V-belt, they rely solely on friction between the belt and sheave surfaces. As a result, flat belts need much greater tension to prevent slippage and the typical tension ratio for good flat belt operation is in

PHOTO 9.1 Showing two sets of banded belts, i.e., belts that share a common outer fabric layer. The set on the left is a pair of B belts while the set on the right shows a pair of 3V belts, that has almost as much capacity as the B belt set.

the order of 2 to 1, with the resultant bearing and shaft loads being much higher than those of a comparable V-belt drive with a tension ratio of 4 to 1 or more.

Synchronous belts, with their teeth fitting into the drive and driven sprockets, usually don't require as much tension as either V-belts or flat belts. Nevertheless, because there is the positive engagement (and the danger of tooth jumping) and because the high performance synchronous belt tension members have some initial stretch, belt vendors generally recommend the total tension in a synchronous drive be the same as the equivalent V-belt drive. Moreover:

- Once the initial stretch has occurred, the tension can be cut in half. (So, when a synchronous belt is reinstalled, the required tension is half the initial.)
- With the lack of continuing stretch, unlike V-belts, synchronous belts don't require periodic retensioning.

BELT DRIVE DESIGN

Sizing a V-belt is not terribly complicated and most manufacturers have web design programs that are readily available and catalogs with the design steps clearly outlined. The major steps in designing a V-belt drive that should last 25,000 hours are:

1. Start with the motor (prime mover) horsepower and rpm. Based on that, select the basic belt size.
2. Look up a service factor (SF_m) from a chart that lists the prime mover design, the type of application, and the number of hours per day. (An easy starting smooth load like a fan would have a lower SF_m than a log chipper that might start up with a heavy load and may see huge load variations during operation. In addition, a unit that only runs occasionally would have a lower SF_m than one that is expected to run 24 hours per day.) Multiply the horsepower from Step 2 by SF_m.
3. Next, select the reduction ratio, then look at the chart for the manufacturer's available sheave sizes. Select the belt length and the center-to-center spacing and from that, find the application factor (AF).
   - The size of the sheave is important because the larger sheave requires less total force and the center distance of the sheaves affects the number of fatigue cycles.
   - In determining this application factor, consider the contact angle. This is the distance the belt wraps around the smaller sheave or sprocket. As the wrap falls substantially below 180°, or a given number of teeth for a synchronous belt, the horsepower capacity has to be downrated.
4. Multiply the horsepower from Step 2 (prime mover hp × SF_m) times the application factor (AF) to get the design horsepower. Divide the design horsepower by the horsepower capacity of the belt for the number of belts required.

The selection process for a synchronous belt drive is similar except that the drive life will be about 12,000 hours. To reach a design life of 25,000 hours with a synchronous belt drive, multiply the SF_m by 1.4.

As shown in Figure 9.3, even though the V-belt design life is 25,000 hours, because of manufacturing, operational, installation, and maintenance differences there are tremendous variations in lives of individual belts.

In the early 1970s, when we called the belt manufacturers and asked them how long a properly designed (according to their catalog) belt application would last, they said 25,000 hours. By the mid-1980s, when asked the same question, they typically said 15,000 to 20,000 hours. In the late 1990s and the early years of the 21st century, the answer to that question was, "It depends". When pressed, they said that the life would "generally be in the range of 15,000 to 18,000 hours", stating that irregularities in operation frequently prevented them from reaching their 25,000-hour goal.

BELT OPERATION DETAILS

Understanding Figure 9.4 is important because it shows the stress cycles the tension member fibers experience as a belt rotates. Starting on the slack or loose side at 1, the belt is deformed as it rotates around the sheave, creating a tensile stress in the fibers. As a point on the belt travels around the driven sheave toward the tight side the tension remains fairly constant until about the midpoint of the rotation when it starts to increase because of the load that has to be accommodated. When the point reaches 2, the full tight-side tension load is realized but the stress from bending around the sheave disappears. Then the belt travels to 3 where, not only does it see the full tight-side tension load, but also the bending load from rotating around a sheave that is smaller than the driven sheave. It then rotates with the sheave until

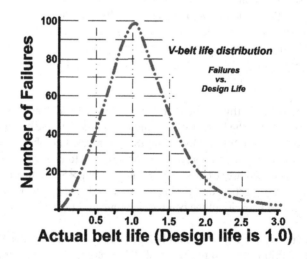

FIGURE 9.3   Showing the actual failure distribution involving over 1000 V-belts in a test lab.

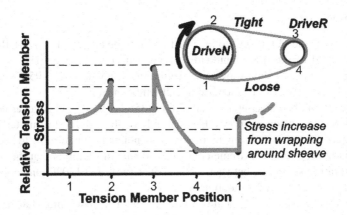

FIGURE 9.4 A diagram of the stress seen by the tension members during a belt rotation.

it reaches 4, where the bending stress disappears and only the slack side tension remains. In this travel sequence around the sheaves the tension member sees two very distinct tension peaks repeated with every cycle and tension member (belt) life is dependent on both the magnitude of the peak values and number of fatigue cycles. The stress cycles seen by a V-belt and a synchronous belt are essentially identical but, because the synchronous belt is thinner, the fatigue stress contribution from bending is much less.

One point that is very obvious from Figure 9.4 is that the use of backside idlers is extremely damaging from a fatigue standpoint. These idlers cause stress reversals in the tension members and the result is shortened belt life.

Some other points are:

1. With V-belts there is always some creep, i.e., differential movement as the belt rotates with the sheave.
2. Serpentine belts, as commonly seen on today's automobiles, have relatively low tension member bending stress because of their thin cross-section.
3. Using larger sheaves results in both lower and fewer fatigue stress cycles.

DRIVE EFFICIENCIES

According to the manufacturer's data and our studies, some typical belt drive efficiencies are:

- V-belts
  a. Just installed and properly tensioned – 97%
  b. Typical V-belts with periodic retensioning – 93% to 94% (Raw-edged belts are typically about 1% +/– more efficient than wrapped belts)
  c. Worn sheaves with loose and flopping V-belts – 88% to 91%
- Synchronous belts – 97+%
- Flat belts – 97% when properly installed, but maintaining correct tension is important.

TEMPERATURE EFFECTS

Temperature is one of the keys to long belt life and the cooler a belt runs, the longer it will last. All of the belt manufacturers have run extensive studies on temperature effects and Gates' data shows that, beginning at about 35°C (~95°F) belt life is halved by every 10°C (18°F) increase in belt temperature.*

The minimum belt start-up temperature is −35°C (−30°F) and is governed by the flexibility of the tension members. But that only applies to the start-up temperature because once the belt is running it will generate enough heat internally to keep it well above this figure, even in extreme cold. A good rule of thumb is that the operating belt temperature should be between 33° and 50°C (90° and 120°F) and above about 60°C (140°F) very rapid degradation occurs.

Elevated temperature degradation is insidious. The first thing that happens is that the elastomer gets softer. This prevents the load from being transmitted evenly across the tension members to the point where the center members are essentially unloaded. Then, with time at temperature, the elastomer ages and hardens until it becomes brick-like. It can't compensate for the normal wear, needs higher tensions to prevent slippage, and when it cracks, doesn't evenly transmit the loads. Adding to the problem is the fact that, as the belt temperature increases the fatigue strength of the tension member fibers decreases. Belt manufacturers say that the major cause of failures is elevated temperature.

BELT OPERATION AND FAILURE CAUSES

A good understanding of the fatigue stresses shown in Figure 9.4 is an important aid to appreciating how and why temperatures affect power transmission belt failures.

Manufacturers say that the top five causes of elevated temperatures in V-belts, in order of their importance, are:

1. Slippage
2. High Ambient Operating Temperatures
3. High Operating Loads
4. Misalignment
5. Belt Bending Stresses

SLIPPAGE

As mentioned earlier, in order for a V-belt to function, there is always some slippage (creep) between the belt and the sheave. When the section of the belt comes into contact with the sheave it wedges itself in, rotates a distance with the sheave, and then has to be pulled out. In an ideal application, the creep (slippage) is about 0.5%, and as the slippage becomes greater, more and more heat is generated.

Slippage isn't difficult to control but, because of wear and tension member elongation, it does require attention on a regular basis. Over the first few hours of operation a belt becomes seated (wears-in) on the sheaves and there is also some minor initial wear of the sheaves. The most rapid tension member stretch and belt tension

---

* www.gates.com/us/en/resources/resource-library

relaxation happens in the hour or two after startup and the rate of stretch rapidly decreases over the first 24 hours. Along with the polyester tension member's slow stretching, the elastomer and the sheave are slowly wearing, so V-belt tension has to be periodically adjusted for good reliability and operating economy. Good practice says the belts should be tensioned at startup, checked about a half hour later, *then retensioned within the first million or so belt cycles.* They should also be retensioned periodically whenever the slippage becomes excessive. (By the time it is announced by squealing the belt life has been tremendously reduced.) How do you determine when slippage is excessive? Check the sheave speeds with a good quality strobe light and when the slippage reaches 4% you know that retensioning should be scheduled (see Figure 9.5).

TENSION RATIOS

Tension ratios are important to V-belt life and, although the term sounds complicated, it is actually fairly simple to understand and to measure. The *tension ratio* is the ratio of the tight side tension divided by the loose side tension during operation.

When the shafts are at rest the tensions are equal. But to get the shafts to rotate, the tight side force has to be more than the loose side force. Comparing these two forces gives us the tension ratio. If the belt is extremely slack and flopping, the slack side force (slack side tension) is very low and the ratio between that force and the tight side force is high. The more tension there is on the slack side, the lower the tension ratio, and many industry experts say the ideal would be to maintain the tension ratio between 5 and 9. If the tension ratio goes lower the force on the bearings increases and the assembly suffers from short belt and bearing life but if the tension ratio is too high the belts slip and fail prematurely. (The ideal flat belt tension ratio is about 2.)

The graph in Figure 9.6 shows the relationship between tension ratio and belt slippage. It shows that, as the ratio grows larger, the slip increases with the result that the belt gets hotter and hotter and the life decreases rapidly. For example, if the drive sheave is 100 mm (4″) in diameter and rotates at 1000 rpm, theoretically a 200 mm (8″) diameter driven sheave should operate at 500 rpm, but because of creep this doesn't happen. Instead, inspecting a very well installed sheave finds it will run at

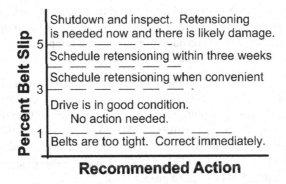

Recommended Action

FIGURE 9.5   This is a guide written for a belt drive that has been in operation for more than one week, i.e., the belts and sheaves have worn in and the initial stretch has occurred. It shows the% belt slip and relates that to a recommended schedule for retensioning the belts.

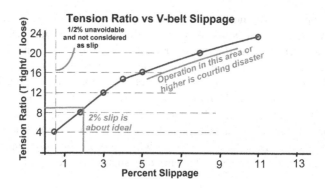

FIGURE 9.6   Tension ration vs. belt slippage, a critical inspection criterion for V-belts.

about 497 rpm while a sloppy, worn, and badly mistensioned application might be turning as slowly as 450 rpm. (The easiest way to determine the percent slippage is to compare the sheave speeds with a strobe light or a digital tachometer. In addition, by using a strobe light and an infrared thermometer, a belt drive can be inspected almost as effectively as a bearing can be inspected with a vibration analyzer.)

One point in understanding the difference in efficiency between synchronous and V-belts is that it takes energy to generate heat and a major advantage of synchronous belts is that they don't slip.

HIGH AMBIENT OPERATING TEMPERATURES

One of the truly frustrating features of many plant operations is the belt guard that looks like it was designed to protect Fort Knox. It isn't uncommon to see one of these guards where it is absolutely impossible to see what is inside – but it's obvious that the belt is hot because the guard is warm to the touch. The designers make these guards to positively guard against anyone putting their fingers into the sheaves but *there is nothing* that says that an intelligent and safe guard can't be fabricated from perforated stock or expanded metal, allowing both cooling airflow and easy inspection.

In a study published by Gates Corporation,* a major belt manufacturer, they found that a 36°F increase in the ambient temperature around a belt reduced the belt life by half. It seems like a relatively simple area to address but frequently we see applications where there is no consideration of ambient temperatures.

In conducting a failure analysis on a belt drive, if the application is in an area that is uncomfortably hot to you, it is also at a temperature level that is reducing the belt life. Take a minute to look at the installation. Could shields be installed? How about cooling vents? Could the shaft have a deflector installed? Should the area be vented? A simple change we have made to several installations is the addition of a fan blade, such as a motor cooling fan, on the shaft between the belt and the source of process heat.

* https://www.gates.com/us/en/resources

High Operating Loads

The more power a belt drive transmits, the harder the drive works and the more heat it will generate. If a well-adjusted belt drive is carrying 75 kW (100 hp) it will have losses of about 4.5 kW (6 hp) and about 3 kW (4 hp) of that is in internal heat from the work, unavoidable slippage, and flexing. On the other hand, if that same drive is transmitting only 37 kW (50 hp) it will generate about 1.1 kW (1.5 hp) in heat. That means the belts on the 37 kW (50 hp) driver will be running much cooler (and will last much longer) that its 75 kW (100 hp) twin.

> How much heat is 1.1 kW? Think about a typical toaster oven that runs at 1500 watts = 1.5 kW.

Belt manufacturers no longer publish factors on the specific effects of load on belt life, but from past data one can see that it is an inverse function close to what is used on roller bearings, i.e., double the load, cut the life by a factor of ten. This brings up the question of loose, mismatched, or missing belts.

- If a drive is missing one of several belts the other belts are more heavily loaded and their life will be reduced by a calculable factor.
- If the belts in the set are all from the same manufacturing lot, they will work in to fit even though one or two seem to be loose on startup. There are tight dimensional controls on the set's manufacturing and, even though some manufacturer's lengths are not absolutely identical, they are within a narrow range. In operation the normal stretch of the polyester tension members will eventually distribute the load evenly across the set.

Misalignment

One of the things every engineer and maintenance mechanic is told is that sheaves have to be "well aligned" or "aligned within 1.5 mm" (~1/16″) or some such data. The reason why those sheaves have to be aligned is fairly simple. Misalignment not only creates additional heat, but it also increases the fatigue stresses in the tension members and show in the following:

1. A V-belt rubs against the side of the sheave groove on its way into position, causing friction that increases the belt temperature.
2. The tension members in the belt are flexed sideward, something that they are not designed to do, and that deformation both creates heat and adds fatigue stress.
3. The misalignment causes vibration that results in additional loads and the harder the belt works the more heat it generates.

Alignment specifications for good reliability vary with the belt type, but regardless, the less the misalignment, the longer the belt life. Looking at manufacturer's specifications shows:

- With individual V-belt drives the two sheaves should be aligned within 4 mm/m (0.06″ per foot) of center distance.
- For banded V-belts, because of their more rigid tension members, the alignment should be within 3 mm/m (0.040″) per foot of C-C distance.
- Synchronous belts, because of the need to mesh the belt and sprocket teeth, require alignment within 2.5 mm/m (0.03″ per foot).

All misalignment reduces belt life.

In addition, before checking the alignment, the sheaves and sprockets should be inspected for excessive runout. Their maximum allowable runout, both axial and radial, varies with diameter but a good rule of thumb is 0.3 mm (0.012″). Again, units with more runout will operate but they will also result in shortened belt life.

One of the problems with these alignment specifications and our normal checking procedure is that we perform a *static* alignment check and, as soon as the machine starts, there is some *dynamic* misalignment introduced. The machine starts and forces are applied to the structure and there will be some distortion. Normally it isn't substantial enough to be critical, but we have seen several failures where that dynamic misalignment has resulted in greatly shortened belt life.

One frequent source of failures is the conversion of V-belt drives to synchronous belt drives without a structural analysis. In most V-belt drives there is some slippage as the drive starts, but the synchronous sprockets don't allow that. They transmit much higher start-up forces and, if the structure deflects and changes the center-to-center distance, the teeth are torn off the belt.

(That 0.3 mm [0.012″] allowable sheave runout is a belt industry standard and pertains to the effect on belt life. But that standard practice doesn't address the damaging effect of the resultant imbalance on the bearing life.)

Sheaves Sizes and Wear

Figure 9.7 gives a clear demonstration of the improvement in life that can result from increasing sheave size. (The peak stresses are developed from a chart like Figure 9.4.) Belts are rated based on the tension member fatigue cycles and reducing the peak load results in huge increases in that life.

The tension members see stresses from both the force needed to turn the sheave and the forces wrapping the belts around the sheave. For example, if the belts used in the drive are transmitting 37 kW (50 hp) at 1780 rpm:

- The force needed to rotate the 300 mm (12″) sheave is 1300 N (292 lbs) and, with a tension ratio of 6, that means the tight side tension is 1575 N (354 lbs) and the loose side tension is 262 N (59 lbs).
- Changing those sheaves to 450 mm (18″), the force needed to rotate the sheaves is reduced to 867 N (197 lbs) and, with the same tension ratio, the tight side tension is 1050 N (236 lbs) and the loose side tension is 173 N (39 lbs).

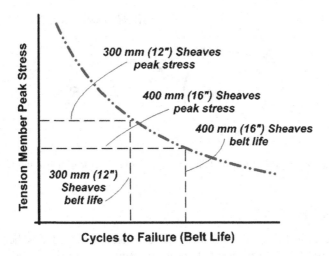

FIGURE 9.7 An example of how larger sheave diameters can result in longer belt life.

With the change in diameter and the resultant reduction in total tension from 1575 to 867 N (354 to 236 lbs), the life of the belt set is increased by more than a factor of 4! In addition, the belts don't have to be adjusted as frequently and bearing life is also increased.

SHEAVE AND SPROCKET WEAR

Manufacturer's data states that worn sheaves can reduce belt life by 50%.* But that only tells part of the story. As the belt wraps around the sheave and the sides bulge outward to grip the sheave, the tension load needed to provide enough friction to prevent slippage has to be increased if the sheave is worn. This additional tension load reduces the belt life and the increased resultant force is then applied to the bearings and their lives are also exponentially reduced. The third negative point about worn sheaves is that they can contribute greatly to a system where the power transmission losses can exceed 10%! Sheaves usually last through three to five belt changes before they have to be changed but the real answer is:

1. Use a sheave gauge and, on conventional V-belts, be sure the wear is less than 0.75 mm (0.030″).
2. If the application uses banded belts the sheave wear should be limited to less than 0.5 mm (0.020″) because more severe wear will cut the belts loose from the band and you won't have a banded belt any more.
3. Synchronous belt sprockets should have no readily visible wear on the teeth.

One final caution about sheave wear is that sheave condition is much more important with raw-edged belts. Wrapped belts have a higher friction coefficient and can tolerate worn sheaves better than raw-edged ones, but excessively worn sheaves will always reduce the efficiency of a belt drive. (Even though sheaves are costly, we have never seen a situation where replacing worn sheaves hasn't rapidly paid for itself in energy savings.)

---

* www.baldor.com/mvc/DownloadCenter/Files/MS4050

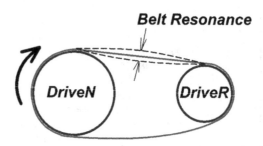

FIGURE 9.8 A diagram showing belt resonance, a key factor in many shortened bearing and belt lives.

BELT RESONANCE

This has become very common with variable speed motors and most belt manufacturers classify it as their biggest field problem. What happens is that, as the speeds, loads, and temperatures vary, the resonant frequency of the belts also varies and it is essentially impossible to design a drive system that doesn't occasionally have the belt resonance as seen in Figure 9.8.

The problem with the resonance is the stress that results in both the belts and the bearings. An easy way of understanding the amount and significance of that stress is to measure the amount of deflection during resonance then, using a belt tensioning gauge, determine the force needed to deflect the belt to that extent. Next, knowing the amount of deflection and the distance between sheaves, calculate the resultant additional operating force on the sheaves. Following through with the design calculations, you'll find the bearing's life reduction to be substantial.

Resonance can affect any belt drive and a simple method of preventing it involves adding a damping device, i.e., a skid plate that just misses the belts during normal operation or another sheave, a kiss idler, that just barely touches them. Then, when the belts reach a resonant frequency, the damping device/idler will prevent the vibration from becoming excessive. *However*, if the skid plate or sheave is installed exactly on the centerline there is the possibility of two harmonics. Better to install it about 40 to 45% across the centerline.

BELT DRIVE FAILURE ANALYSIS TECHNIQUES

Table 9.2 shows the most common causes of belt failures and following it are a diagnostic outline and a series of belt photos showing and explaining some of the common failure mechanisms.

In analyzing a failed belt drive, it would be ideal if there was the ability to check the belt temperature during operation, but we are going to assume that is impossible. Therefore, the suggested procedure is:

1. Rotate the sheaves to measure their axial and radial runout, then check the sheave alignment. Are all the measurements within specifications for the belt type?

TABLE 9.2
Common Belt Failure Causes

SYMPTOM	CAUSE												
	Misalignment	Inadequate Tension	Sheaves Worn or Damaged	Design or Structural Weakness	Overloaded	Wrong Size	Pulsating Loads	High Operating Temperatures	Contamination	Resonance	Installation Damage	Rubbing	Normal end of Life
V and Serpentine Belts													
Belt slippage		XX	XX		X	X			X				
V-belt cover ruptured											XX	X	
Belt turned or jumping	X	X		X			XX			XX			
Elastomer cracked	X	X		X				XX	X				
Bottomed in sheave			XX			X							
Tension member separation	XX		X								XX		X
Synchronous Belts													
Tooth wear	XX				X		X			X			
Belt cover ruptured	X										XX		
Tooth fractures		XX		X			X						X

XX – indicates most likely causes (don't forget to look for multiple causes)
X – indicates a possible cause

PHOTO 9.2 This shows the instantaneous failure of a group of four belts. There is no fraying of the tension members and that tells that the failure happened very rapidly. This is similar to a ductile or brittle overload of a metallic piece and says that the cause of the failure was present *immediately* before the failure occurred.

PHOTO 9.3 The heavy wear on the synchronous belt should have been readily visible during a strobe light inspection anytime in the past month or two. The belt eventually failed as the badly worn teeth were ripped off. Also, the heavily frayed tension members should have given a visible warning of the failure.

2. Inspect the sheave sidewall condition. (Is it smooth and polished or has there been damage, leaving it rough and irregular?) Then, using a sheave gage on V-belt drives or a micrometer on synchronous drives, measure to see if the sheaves/sprockets are in acceptable condition?
3. Compare the sheave size/specification with the belt identification to ensure they are compatible.

4. Examine the broken V-belt.
  a. What do the ends of the tension members look like? Are they unfrayed or only slightly frayed, indicative of a very short-term overload failure as shown in Photo 9.2, or are they severely frayed, showing a long-term failure as seen in Photo 9.3?
  b. Examine the elastomer.
      i. Is it extremely hard with cracks in it, i.e., showing the result of slippage and/or high operating temperatures and degraded elastomer?
      ii. Does the elastomer feel soft or greasy? (This is Indicative of foreign material contamination from grease, oil, chemicals, belt dressing, etc.)
      iii. Are there some random areas that appear charred and worn? If so, suspect spin burns (see Photo 9.12), i.e., places where the belt stopped for a short time while the sheaves continued turning – with the result that the belt became severely overheated.
  c. Do the sidewalls show polishing? Indicative of slippage.
  d. Is the cover separated? Indicative of poor installation practices (Photos 9.4 and 9.5).

Photos 9.6–9.15 show a series of examples of belt failures and causes.

PHOTO 9.4   The cover separation on near belt on the right side of the photo is obvious. This is the result of improper installation, likely either rolling the belt onto the sheave or prying it on with a hard-edged tool.

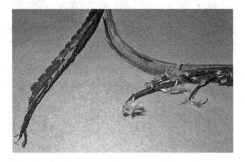

PHOTO 9.5   The separated cover leads us to suspect another installation problem, such as rolling the belt onto the sheave, although it may have been a manufacturing defect. The badly frayed tension members indicate that the failure didn't happen instantly and the belt has been running with a very visible defect for a while.

PHOTO 9.6   This is a pile of belts that has just been removed from a fan. The elasto-mer on the sheave contact surfaces is shiny and polished, proof that they have been slipping excessively.

PHOTO 9.7   The underlying construction fabric on the right side of these synchronous belt teeth is showing and the uneven wear is obvious evidence of misalignment.

PHOTO 9.8 The worn teeth on this synchronous belt are breaking off and it has worn through to the tension members, proof that the belt has reached the end of its design life. It shows some indication of harder contact on the right side and that leads to slightly suspect alignment.

PHOTO 9.9 This is spin burn damage on a small belt. In this case the machine jammed and the belt tension was so loose the drive sheave slipped and heated the belt until it charred. This damage is common when a machine jams but it can also happen from lack of tensioning. During operation a belt constantly stretches and, on a machine such as a fan, when the tension ratio goes over about 16 (5% slip) the chance of a catastrophic failure increases rapidly.

PHOTO 9.10   A sealed belt guard with the resultant trapped heat helping shorten the belt life. In addition, this construction effectively eliminates the ability to inspect the belts in operation. (But a strobe light on the shafts could be used to calculate the slip ratio.)

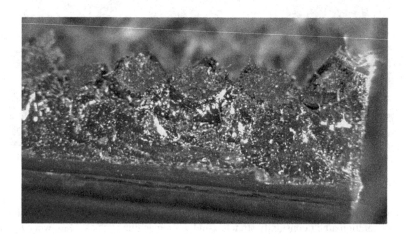

PHOTO 9.11   This photo was taken though a microscope and shows a section of a synchronous belt. Along the top can be seen the round tension member cords and the fact that they are broken off cleanly indicates that the belt failure was either the result of a shock loaded or being pinched.

PHOTO 9.12 A heavily worn synchronous belt sprocket that should be replaced, but note the broken belt. The ripped backing caught our eyes and further inspection found the tooth fractures started at the center of the belt. The most likely cause of the belt failure was the combination of gross misalignment and rolling the belt onto the sprocket, damaging the backing and tension members.

PHOTO 9.13 Look at the surface of the top cover on this synchronous belt!! The foreign object imprints have caused stress concentrations that will significantly reduce the life of the tension members.

PHOTO 9.14 This is a belt that has been removed from service. The cracks in the elastomer show that it has become hard and brittle and the belt should be replaced. But, look closely at the pattern on the side of the belts. The different surface appearances show that it has been used in the wrong sheave group, i.e., this is a modern V-belt that has been used in classical V-belt sheave.

PHOTO 9.15 This doesn't show a failed belt but is a great lesson on what a belt drive shouldn't look like. The additional load on the bearing and shaft from the sheave being installed this far away from the bearing is ridiculous. If the design were better, the sheave could have been moved in by at least 4 in (100 mm) and the bearing life would be at least tripled.

10 Ball and Roller Bearings

The first documented use of roller bearings that we could find was in 300 BC when the Greek engineer, Diades, developed a roller-supported battering ram. Another early application can be seen in the Danish National Museum where there is a series of rollers that are over 2000 years old.

In the 17th and 18th centuries, there were types of rolling element bearings being used for applications, such as windmill center posts, but the problem with these bearings was that at the time they didn't have the manufacturing techniques to make the balls (or rollers) exactly the same diameter. The result was that some balls were much more heavily stressed than the others and frequent failures occurred. (This goes back to the heart of understanding rolling element bearing life. The effect of increased stress [load] is to reduce the element life by at least a cubic factor. So if one element takes a load much greater than its share, that element will rapidly fail and the bearing's life will be dramatically reduced.)

Modern rolling element bearings began in 1886 in Germany when Friedrich Fischer designed and built the first machinery for the production of precision bearing balls. Other important early developments included Henry Timken's invention of the tapered roller bearing in Missouri (United States) in 1898, and Sven Winquist's invention of the first self-aligning ball bearings in 1907 in Sweden.

PARTS OF A BEARING

To ensure we're all thinking along the same lines, Photo 10.1 shows and identifies the parts of a typical ball bearing. One point we should make is that the pieces we call the inner and outer rings, many people in North American industry will identify as the inner and outer races. For many years, the term "ring" has been used in the rest of the world and there has been an agreement within the bearing industry to standardize terminology across the world and we use the term "ring" because it agrees with them.

BEARING MATERIALS

To withstand the extremely high Hertzian fatigue stresses and give relatively long life, the steel used in rolling element bearings is heat treated to be very strong and hard. Most bearing rings and elements are through hardened and are made from SAE 52100 or a similar alloy, and every major country has their own standard, such as Japan's SUJ2 steel.

The mass of a part affects the heat treating response, i.e., the chemistry of the steel can change a little depending on size of the bearing, and some larger bearings use a slightly different alloy, but the difference in the through hardened alloys isn't substantial.

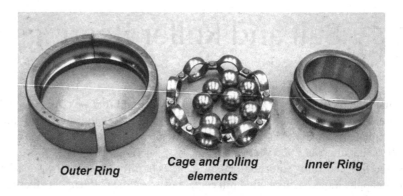

PHOTO 10.1   The components of a ball bearing.

However, there are also surface hardened (case hardened) bearings and they are most commonly made from either AISI grade 4118H or 4320H or their European and Japanese counterparts. In addition, there are some relatively uncommon bearing steels that have other additions to improve their toughness and fatigue resistance and there are also a number of ceramics and unusual alloys that are being used for rolling elements and as coatings.

The composition of the through hardened steels varies a little in the amount of silicon and manganese while surface hardened ones generally vary the amount of nickel and chrome. These variations are relatively small and are not critical to the user but a good understanding of the differences between the two basic structures is important and Figure 10.1 shows them.

Surface hardened bearings are more costly, and in most applications, there is not a substantial performance difference between the two. However, in applications where ring cracking can occur, mainly those where shock loading is present, the

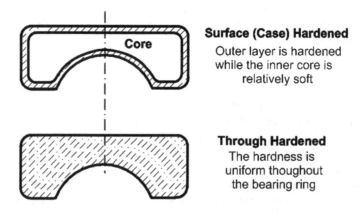

FIGURE 10.1   Showing the difference between through and surface hardened bearing rings.

surface hardened units have an advantage because the cracks don't readily propagate through the softer core and across the entire ring.

Photos 10.2 and 10.3 compare the damage in a case hardened tapered roller bearing with that of a through hardened spherical roller bearing. Both have cracked rings but the two-year-old tapered roller bearing is still in somewhat usable condition, because the crack is has not propagated through the core. Meanwhile, the two-week-old spherical bearing has cracked all the way through and is only a short time from a catastrophic failure. Common situations where case hardened bearings are used include paper machine dryer bearings that are subjected to thermal shocks and logging and mining equipment that experiences impact loading.

PHOTO 10.2   This is a case (surface) hardened bearing that has been running cracked for almost a year without catastrophic failure.

PHOTO 10.3   This is a through hardened bearing only a few days from disaster.

When metals become harder and stronger, their fatigue life increases and, although the rolling elements are usually a bit harder than the rings, the surface hardness of bearing components generally ranges between HRC 58 and 62. Reducing this hardness reduces the fatigue strength (life) of the bearing and a long-used rule of thumb is that for every point the hardness is reduced below HRC 58 the bearing life is cut by about 10%.

Why did Harry Timken start making his tapered roller bearings case hardened instead of through hardened? Because! We've talked with a lot of people and asked them the same question without a definitive answer. However, the consensus is that the common method of hardening components in Henry Timken's world, St. Louis, Missouri, in the 1890s, was pack carburizing and that was the process he used, while in Germany, it was common to through harden materials.

The normal maximum operating temperature where conventional bearing steels can be used for long periods of time without a reduction in capacity or reliability is about 150°C (~300°F), while the M50 and 440C bearings can be regularly run as high as 220°C (425°F). (In both cases, the steels doesn't actually begin to weaken for more than another 50°C [90°F] but the bearing operating stresses will create additional heat and it is wise to leave a reasonable safety margin.) If you are talking about very short term, a few minutes of heating, of an SAE 52100 bearing, 175°C (350°F) can be tolerated without reducing the life but bear in mind that any seals or lubricants exposed to temperatures over about 120°C (250°F) will likely be destroyed.

Valuable for many applications, several very effective coatings have been developed to improve life in wet, corrosive, and contaminated operations. Among others, these include Torrington's TDC (thin dense chrome-coated) bearings and Timken's Aqua Spex. These coatings have been applied to the rolling elements, rings and cages. They are one of the areas of a great deal of current research and there will be new developments concerning them. (Our experience with TDC bearings is that, compared with a standard bearing, a four-fold increase in life in not unusual in wet and dirty applications.)

There are some specialty bearing materials used for corrosive, high temperature, and exotic services but they usually are not seen in general industrial applications. Stainless steels, such as AISI M50 tool steel and AISI Type 440 C stainless steel, polyamide, ceramics, and nylon, are all used for these applications. One increasing use of ceramics such as silicon nitride is the rolling elements of bearings designed for electrical isolation of the inner and outer rings. They are also being used with success in variable speed and many high speed motor applications.

CAGES

The functions of the cage are to maintain the proper spacing between the rolling elements, to minimize friction, and to distribute the rolling elements' load carrying ability evenly around the rings. There are dozens of cage materials and the proper selection can be critical to both the bearing's performance and the cost.

For many years the most common cages were made from stamped steel with the two halves riveted together. (This is called a Conrad bearing in honor of Robert Conrad, the British engineer who developed the machine to assemble the cages.) Interesting is that it has been discovered that, in many grease lubricated applications, the cage carries a substantial portion of the lubricant that resupplies the rolling elements during operation.

The classical advantages and disadvantages of some cage materials are:

- *Stamped and Riveted Steel* – Low cost, poor frictional characteristics when there is marginal lubrication, good thermal expansion compatibility with steel bearing components, relatively weak and deformable.
- *Stamped one-piece steel* – Inexpensive, used where stresses are higher particularly in roller bearings.
- *Bronze* – Excellent natural lubricity for applications where lubricant is marginal, costly. Stamped cages for moderate stresses while cast and machined cages are frequently used where cage stresses are high, such as with nonuniform rotation or high vibration.
- *Polymers* – Light weight, self-lubricating, easily produced with good corrosion resistance. Polymer cages are frequently a composite of two or more materials with the properties varying depending on the specific materials used. Common cage materials are nylon, polyethylene, polyamide, etc. with the advantage that they can be formulated to be strong, have good water resistance, or high hardness, or high strength at elevated temperatures as needed.

BEARING RATINGS AND EQUIPMENT DESIGN

To determine the life of a ball or roller bearing, the bearing companies initially used a huge sample of identical bearings, all lubricated in the same manner and subjected to the same loads. These bearings were then run selected speeds with controlled temperatures and ideal lubrication, and detailed records were kept of their failure rates. Combining the results from many similar experiments, enabled the development of the Lundgren-Palmgren equations for rating bearing lives and, in most cases, the bearing dynamic rating is specified as the rated load where 90% of the bearings will run for more than one million revolutions.

The bell curve in Figure 10.2 shows the life distribution that resulted from these experiments. On the curve are marked the L_{10} life, the point where 10% of the bearings have failed, and the L_{50} life, the average bearing life. It is interesting to note that the curve is distorted a little to the right and there are some bearings that seem to last forever.

From a practical standpoint, important points to understand are:

1. Most equipment designs are based on the L_{10} life of the bearings.
2. The L_{10} life is determined by these experiments and by the Lundgren-Palmgren equations and is based on failure by Hertzian fatigue.

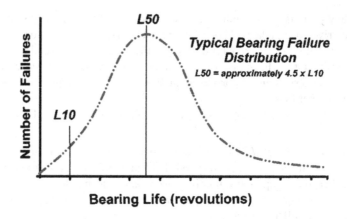

Bearing Life (revolutions)

FIGURE 10.2   This shows the typical expected distribution of bearing failures.

3. The average bearing life, the L_{50}, is the point where 50% have failed and for most bearings is about 4.5 times the L_{10} life.

But what we have found in the real world is that contamination and poor lubrication will lead to surface fatigue failures and those bearing lives will be much shorter than the calculated Hertzian fatigue life.

The typical designed L_{10} lives for some common machinery are as discussed in the following paragraphs. (Note that these are not the actual lives; see Table 10.1.)

Using these guidelines, if we were to select a rear wheel bearing for a heavy duty pickup truck, we would first determine the expected loads and then go to a manufacturer's bearing selection data and do the calculations for a bearing that would fit the bearing housing and have an L_{10} life of at least 360,000 km (220,000 miles).

TABLE 10.1

Some Common Bearing Design Lives

Equipment Type	Typical L_{10} Life
Domestic hand tools	1500 hours
Domestic motors	1500 hrs
Small industrial motors (less than 30 kW[40 hp])	20,000 hrs
Large electrical motors (over 30 kW [40 hp])	50,000 hrs
Industrial centrifugal compressors	80,000 hrs
Fans	
General industrial	40,000 hrs.
Mine ventilation	100,000 hrs
Industrial reducers	Depends on service factor
Pickup trucks – light duty	250,000 km (150,000 mi)
Pickup trucks – heavy duty	360,000 km (220,000 mi)

Understanding the effects of changing loads on a rolling element bearing is absolutely critical to realizing good reliability. For a ball bearing, the bearing life is rated using the formula:

$$L_{10}(\text{life}) = [\text{Basic Dynamic Load Rating} / \text{Dynamic Equivalent Load}]^3 \times 10^6 \, \text{rev}$$

Therefore, from the formula, we see that doubling the load on a ball bearing will cut the Hertzian fatigue life by a factor of 8. In a similar manner, the exponent used for roller bearing life calculations is 3.3 and doubling the load on a roller bearing will cut the life by a factor of 10!

If the load changes, the life will exponentially increase or decrease. For example, if the load on that ball bearing were to be doubled, to 908 kg (2000 lbs) the L_{10} life would be reduced to 11.7 million cycles. That sounds like a lot of revolutions but, when you consider that an 1800 rpm motor turns more than 2,500,000 rev/day, that isn't even a six-day run!

The Hertzian fatigue life is very sensitive to loads and, in another example, let's select a ball bearing that has a BDR of 454 kg (1000 lbs) and a load of 50 kg (110 lbs). Then, according to the table, dividing 50 kg by 454 kg=0.11 and, the bearing has a normal load rating. Next, to calculate the L_{10} life we would use the standard industry formula

$$L_{10} = [\text{BDR} / \text{actual load}]^3 \times 1,000,000$$

substituting

$$L_{10} = [454 / 50]^3 \times 1,000,000 = 748,000,000 \, \text{cycles}$$

Looking at common designs and their design load classification in Table 10.2, we see:

TABLE 10.2

Bearing Loads and Classification

Bearing Type	Bearing Load Rating		
	Light	Normal	Heavy
Ball	<0.07 BDR	0.07 to 0.149 BDR	> 0.15 BDR
Roller	<0.08 BDR	0.08 to 0.179 BDR	> 0,18 BDR

BDR – Basic Dynamic Load Rating

If the machine runs all day every day at 1800 rpm the L_{10} life would be about 9.5 months. From Figure 10.2 we know the average bearing life is about $4.5 \times L_{10}$ so the average bearing, the L_{50}, should last about 3.5 years.

If the load changes, the *life will exponentially increase or decrease*. For example, if the load on that ball bearing were to be doubled, to 100 kg (220 lbs) the L_{10} life calculation shows

$$L_{10} = \left[454 / 100\right]^3 \times 1,000,000 = 93,500,000 \text{ cycles}$$

The net effect is that increase in load cut the L_{10} life from over nine months to just a little over a month!

As bearing steel manufacturing practices improve bearing life will improve, but it is important to realize that there will always be some bearings that fail from Hertzian fatigue before the L_{10} is reached, and there are many applications where the failure of 10% of the bearings is not acceptable. For example, if you have twenty bearings in a system, it is almost a certainty that at least one will fail before the L_{10} life is reached. In those applications where greater reliability is needed the design criteria might be an L_1 or even an $L_{0.5}$ and the effect would be to specify larger bearings as is shown in Table 10.3 to the right.

HERTZIAN FATIGUE, ROLLING ELEMENT BEARING LUBRICATION, AND SURFACE FATIGUE

We mentioned earlier that all basic bearing designs and these formulas are based on the assumption that the bearing will fail from *Hertzian fatigue*. (Heinrich Hertz, the same person as the electrical frequency is named after, developed the initial equations for expressing the fatigue stresses experienced by a flat plate exposed to a load from a rolling element about 150 years ago and all rolling element bearing design calculations are based on the further development of his work.)

Figure 10.3 shows a typical ball or roller as it goes along its path. Looking at this diagram, one can see that, as the rolling element proceeds along the ring, the surface elastically depresses in reaction to the compressive force and this depression results in *Hertzian stresses* within the part. Fatigue failures cannot occur with purely

TABLE 10.3

Multiplier for Improved Reliability

Desired Reliability Level	BDR Multiplier
L_{10}	1.0
L_5	1.6
L_4	1.9
L_3	2.3
L_2	3.0
L_1	4.8

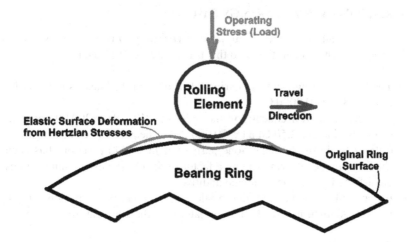

FIGURE 10.3 This sketch shows the elastic deformation that allows Hertzian fatigue to occur.

compressive loads but what happens within the bearing ring is that the areas immediately before and after the depression spring upward and are internally subjected to tension stresses. It is the tension in these areas, followed by the compression from the rolling element, that causes the Hertzian fatigue cracking.

A Hertzian fatigue failure begins with a fatigue crack that grows parallel to and about 0.1 mm (0.004″) below the surface, and then rapidly works its way to the surface. However, in reality most of the industrial bearing failures we've seen have been from surface fatigue, cracking that starts on the rolling surface and is aggravated by contamination, dimensional errors, and poor lubrication. Figure 10.4 shows the difference between how surface fatigue cracking and Hertzian fatigue cracking begin.

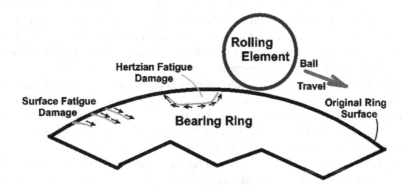

FIGURE 10.4 The difference between the common surface fatigue and Hertzian fatigue, the designed failure mode.

THE REASONS WHY BEARINGS FAIL

In the chapter on lubrication we discussed how *viscosity conversion* works to support the bearing element in operation, and this is a good time to reiterate:

- The typical rolling element bearing lubricant film thickness is in the order of 0.0008 mm (0.00003").
- The pressures are frequently in the order of 1.40 GPa ($\approx$200,000 psi) but can go well above 2.50 GPa ($\approx$370,000 psi) in heavily loaded bearings.
- 0.1% water in the lubricant (one part water in 10,000 parts oil) has been shown to reduce bearing life by a factor of 5. (But the exact damage rates are affected by the lubricant and additives.)
- It's generally accepted that the smallest particle the typical person can see is about 38 microns (0.0015"), far greater than the lubricant film thickness.

According to bearing company data, most bearings are replaced because of wear but it is uncommon that a failure analyses would be conducted on a worn rolling element bearing. Instead, we find badly damaged and occasionally melted pieces of metal and often it's hard to find all of the pieces. From our analyses, it appears that the major causes of bearing problems result from ignorance of the four statements above and below, and we'll look at them in more detail.

The typical rolling element bearing lubricant film thickness is in the order of 0.0008 mm (0.00003"). The film thickness will vary depending on the relative speed, the lubricant viscosity, and the surface roughness, but what is important is that the lubricant film acts to spread the load over a greater area and reduce the stress concentrations in the element path.

Critical is that dirt particles smaller than we can see and larger than the film thickness can damage both the ring and element surfaces and will act to interrupt the film. The effect of these particles is to cause "bruising", denting of the elements and rings, and increase the local fatigue stresses and ISO 281-2007 and several bearing manufacturer's texts give guidance as to exactly how this contamination can reduce bearing life.

The normal operating pressures are tremendous and are frequently in the order of 1.4 GPa ($\approx$200,000 psi) but can go above 2.5 GPa ($\approx$370,000 psi) in very heavily loaded bearings. If we add to that the stress from an out-of-round bore, or a bore that doesn't properly support the shaft, and/or the vibration forces from a loose shaft fit, the life of the bearing can be tremendously reduced.

Table 10.4 is a listing of typical internal clearances for deep groove type ball bearings. Reviewing it, one can see that, for a normal bearing with a 50 mm (~2") bore the internal clearance can range from 8 to 28 μm (0.00035" to 0.0011"). The machining specifications for the bore the bearing fits into call for the maximum allowable out-of-roundness to be less than 14 μm (~0.0005").

From the table it is easy to understand how a 0.5 mm (0.020") "soft foot" or the deflection of a weak base could distort the bearing housing and the outer ring and greatly reduce the bearing life.

TABLE 10.4
Ball Bearing Radial Internal Clearances

Bore Range		Typical Internal Clearance					
		Metric min/max in µm			Inch min/max in 0.0001″		
Over	Up to	Normal	C3	C4	Normal	C3	C4
30	39	6/20	15/33	28/46	2/8	6/13	11/18
40	49	6/23	18/36	30/51	2.5/9	7/14	12/20
50	64	8/28	23/43	38/61	3.5/11	9/17	15/24
65	79	10/30	25/51	46/71	4/12	10/20	18/28
80	100	12/36	30/58	53/84	4.5/14	12/23	21/33

The distortion caused by "soft foot" should not be underestimated. Our company did the predictive maintenance for a plant that had repeated failures of some inexpensive cast iron, two bolt pillow blocks. Analyzing the ball paths of the failed bearings, we found there were two load zones, one centered at about four o'clock and one at about eight o'clock. Our initial reaction was to think the person didn't properly shim the pillow block but we eventually realized that the inexpensive fan base was deflecting under load, causing distortion of the housing, and we revised the base to include a 12 mm (1/2″) top plate. That plant had six fans with similar bases and, since then, more than 30% of the two-bolt pillow blocks we've looked at have shown bore distortion problems. Some of that distortion is from sloppy assembly, some is from a lack of manufacturing flatness standards, but much of it is the result of "dynamic soft foot" where the operating stresses cause structural distortion.

Like many of the other failures we've mentioned, we frequently see more than one cause and it is not unusual to see an application with an incorrectly installed bearing with contaminated lubricant in a loose fit – and the plant personnel are blaming the machine manufacturer for the short bearing life!

A DETAILED ROLLING ELEMENT BEARING FAILURE ANALYSIS PROCEDURE

The procedure below uses nine basic steps to be followed in looking for the causes of a failed bearing and within each of the steps are numerous questions to be asked. It starts with an inspection of the housing and surroundings, to better understand the bearing's operating conditions, then looks at the exterior of the bearing (to understand the heat transfer and supporting forces), then the internals (to see the actual operating forces).

The nine steps we use in finding the physical causes of a bearing failure are:

1. Understand the background, i.e., the bearing's history, the design criteria, and the plant's lubrication practices.
2. Inspect the shaft and housing contact surfaces of the inner and outer rings, i.e., the inner and outer ring fits.
3. Next, look at the sides of both ring exteriors.
4. Remove the seals and/or shields and visually examine the lubricant.
5. Look at the application, determine the loads, understand what the ball and roller paths should look like, and separate the rings so those paths can be inspected. Then place the rotating ring on a lazy Susan and slowly rotate the bearing while watching the ball/roller path.
6. Continuing with the physical inspection, place the fixed ring on a lazy Susan and slowly rotate it while watching the ball/roller path.
7. Hardness test the rings and rolling elements.
8. Inspect the cage paying careful attention to the element contact surfaces, looking for indications of wear and/or misalignment.
9. Inspect the rolling elements.

Below we've added a group of questions to the nine points that, when they are answered, should lead to the multiple roots of the failure. In listing these questions, we've tried to make comments about the various answers and why one or the other may be of importance. We've also included photographs illustrating many of the points. Please realize that there will be times when some of the answers won't necessarily point to the bearing problems. They are used to help the investigator develop a better sense of the operation. However deviations from what is normal and desirable are clues to failure causes and a starting point for further investigation.

Reading these steps will be confusing the first time around because there is lots of text and there are many questions. When that happens, go back and look at the list of the nine questions and try to patiently think about what you have seen. What we have tried to do is write down how we go about looking at a failure and how to interpret what is seen. Like the shaft failures mentioned earlier, the parts will tell you how they died. The difficult part is interpretation of the symptoms but with patience and lots of practice it becomes much easier.

1. Understand the background, i.e., the bearing's history, the design criteria, and the plant's lubrication practices.
   1) *When was the equipment installed? Was this a new installation?* Has the machine loading or application changed from when it was first installed? (What effect should the increased load have on the life?) What has the historical replacement schedule been? (Is this typical of past failures?) Was this bearing installed at the factory or is it a replacement? (What is the likelihood of a human error in a field assembly?) If

the bearing is out of a pillow block or a machine like a motor or reducer, check for machine base distortion, i.e., soft foot. (An inadequate base will cause distortion of the bearing's outer ring and increased internal loads.)

2) *Look at the housing and the bearing's surroundings. Are they clean and uncontaminated?* Some thoughts about the surroundings are:

   a) Bearings are not perfect and they generate heat. If debris is covering the housing it will increase the temperature of the parts inside. In turn, increased internal temperatures will lead to reduced lubricant viscosity, reduced film thickness, and more rapid additive deterioration. The bearings and gears in the conveyor drive in Photo 10.4 are being destroyed by their environment.

   b) Water in lubricants is deadly to bearing life and if the housing has been wet there is the possibility that the lubricant has been contaminated.

3) *What is the range of temperatures and relative humidity readings in the area? Is the shaft hotter than the bearing?*

   a) If the shaft is hotter than the bearing, the inner ring is going to grow much more than the outer ring and problems outlined in 1 above could occur. If this is the case, what internal clearance is specified? (Motor bearings are usually C3 clearances because the shaft runs hot. C4 bearings have even more internal clearance.)

   b) The comment about relative humidity is directed at the possible effects of water contamination in the lubricant.

4) *When was the bearing installed? How fast does it rotate? Is the bearing application correct for the expected speeds and loads.? How long has it run? Are the operating conditions still the same as the machine was designed for?* These questions go back to the design of the bearing and its application – so the analyst has a good idea of the design conditions

PHOTO 10.4   A conveyor drive covered in insulating dust on a deformed base, two conditions that guarantee poor reliability.

and can compare them with the operating conditions. Two things to keep in mind at this stage of the analysis are:

a) Ball bearing life is inversely proportional to the load to the 3rd power while for roller bearing life the factor is 3.3.

b) Bearing design is for a given number of revolutions and the effect of changing speed is that those design revolutions will be used up faster or slower.

c) There are specified limiting speeds for bearings and they will depend on the lubricant. However, if the application exceeds the maximum speed, most of the bearing manufacturers do have alternative surface finishes that allow higher speeds.

d) Is this the correct bearing for the installation?

5) *What is the lubricant, what is the relubrication schedule, how is the relubrication accomplished, and what education have the oilers had?*

a) Is the bearing being lubricated with grease or an oil? (An advantage of oil lubrication is that it can do an excellent job of controlling the bearing's temperature.)

b) When you look at how often the unit is being lubricated, is the real purpose of that relubrication to resupply the rolling elements with lubricant or is grease being used as an auxiliary seal?

c) How much grease is being pumped into the bearing and how is it applied? (That tube of grease is about 80% oil and the thickener's job is to act as a reservoir and let the oil slowly wick out and lubricated the rolling elements.) Our inspection data found that a substantial percentage of bearing failures are hastened by overlubrication and Chapter 8 has a relubrication guide that can be used to determine the relubrication frequency and lubricant quantity for most bearing applications above 100 rpm.

Photo 10.5 shows a shielded motor bearing that has been destroyed by a zealous but untrained oiler. Shielded bearings can be relubricated if a measured quantity of grease is slowly and carefully pumped in by hand but destroyed bearings with dimpled shields like these can only be produced by an uneducated finger on the trigger of an automatic grease gun.

Continuing on the thought of proper lubrication, although this photo shows a shielded bearing where the failure was caused by poor practices, damage can also be caused by overlubrication of open bearings. *It might be hard to think that too much grease can cause a problem but, on one of our test stands, we intentionally overgreased a ball bearing, and the housing temperature went up by more than 55°C (100°F) in less than 10 minutes. Think what that did inside the bearing!* With overlubrication, the steps in the typical failure involve:

a. The oiler pumps excessive grease into the bearing cavity.

b. Then the rolling elements have to continually push the lubricant out of the way and that generates more heat than normal.

PHOTO 10.5   A shielded bearing where the shield has been collapsed onto the cage and balls by careless greasing.

      c. The excessive heat not only causes reduced internal clearances and increased parasitic loads but also it reduces the lubricant viscosity and film thickness.

      d. The result of the excessive loads and reduced film thickness is greatly reduced bearing life.

  6) *Review the maintenance records, including the vibration analysis and/ or oil analysis history on the machine.* In analyzing this data, is there a pattern showing a relationship between the first measurable deterioration and other plant activities?

  7) *Speak with the operating, maintenance, and engineering personnel involved with the machine. Ask for their ideas on what could have contributed to the problem.*

2. Inspect the contact surfaces of the inner and outer rings, i.e., the inner and outer ring fits, and answer the following questions.

  1) *Measure the bearing housing bore and the shaft dimensions in at least three positions on each side of the contact. Compare these readings with the specifications.* After the basic lubrication is supplied, probably the most important single characteristic of a bearing installation is the dimensional fit. The bearing generates heat and, unless the bearing is in an oil bath, most of that heat is usually conducted out through the shaft and housing fits. If those fits are not sound the internal bearing temperature increases and the clearances decrease. (Most of the heat is generated by friction but the elastic flexure of the bearing steel also generates some heat.) The heat generated is approximately equal at both rings but the inner ring is almost always smaller and has less mass, so it gets hotter and grows more. When the clearances decrease, the internal stresses increase. In addition, as mentioned earlier, the increased temperature reduces the viscosity of the lubricant and when viscosity conversion takes place the resultant film is thinner and the fatigue stress in the rings is increased.

2) *Inspect the bearing surfaces*
   A. Are there any black stains on the rings?
       Yes – There has been water present.
       No – Water has not contaminated the bearing.
       Comment – We know the critical effect water has on bearing life. The water contamination typically causes black stains on the rings or in the ball path as shown in Photo 10.6. The photo shows the actual ball path damage and the presence of stains such as these or the ones shown in Photo 10.7 are proof that water has been present.

       The first time I really recognized the importance of those "black spots" caused by water was early in my career when we were trying to understand why the most expensive bearings in the plant were failing at a greatly increased rate. One day we were inspecting one of the failed bearings with a rep from the bearing company when he pointed out the black corrosion spots. They didn't look or feel like much and we asked if we couldn't just take fine emery cloth and polish them out. The

PHOTO 10.6   Two water spots (corrosion locations) on the inner ring of a ball bearing.

PHOTO 10.7   The side of a bearing and those black spots are also the result of corrosion and proof that water has been present.

rep went on to say we could do that but it wouldn't help the situation because the actual mechanism is:

1. Water gets into the bearing and there is some corrosion.
2. The corrosion generates atomic hydrogen and some of the hydrogen atoms are absorbed into the steel.
3. The hydrogen atoms wander through the steel until they find another hydrogen atom and the two unite to form a hydrogen molecule.
4. The volume of the molecule is much greater than that of the atom and it results in enough internal pressure to cause cracking of the extremely brittle bearing ring.

If we were to clean off the corrosion debris, that black oxide, it wouldn't do any good because the bearing steel already had tiny cracks in it. He went on to say that if we did run the bearing, he felt we would be lucky if it lasted for one-tenth of the L_{10} life.

B. *Other than water stains, is there any visible polishing, fretting or discoloration of the shaft or housing fit surfaces?*

Yes – The bearing has been moving in the housing or on the shaft.

No – The bearing fit was excellent.

Photo 10.8 shows a bearing where the design called for only one half to be supported. From it we can see the difference between a what a loose fit and a correct one should look like. The major problem with the loose fit is the reduced heat transfer.

Comment – Recognizing fretting is important because it is proof that the bearing has been moving in the housing. Sometimes this movement is allowable, e.g., the floating bearing in a motor where having two fixed bearings would lead to very short bearing life. But it is important to look at the installation and recognize if fretting has occurred.

Photo 10.9 is an example of *fretting*. This bearing was loose in the housing and moved back and forth a tiny bit as the inner ring rotated. The brown-black material is the oxide that the movement created.

PHOTO 10.8 The upper portion of this bearing shows the polished surface typical of a loose fit. The bottom half is what a good fit should look like when a bearing is removed from use.

PHOTO 10.9 The dark material on the outer surface of this ring is fretting corrosion and results from looseness between the ring and the mating surface. It is actually a collection of metal particles that have been abraded off the surface and then oxidized into rust.

    C. *Is the original grinding pattern still visible over more than 50% of the surface?*

        Yes – Some fretting is acceptable. If most of the original bearing surface isn't attacked, the bearing ring has been moving but the fit is acceptable.

        No – The movement is excessive, causing fretting and reducing the bearing's life.

    Comment – As we mentioned earlier, there are times when some fretting is inevitable. As long as it isn't excessive the insulating effect is not substantial. If most of the original ground outer surface is visible the fretting wasn't excessive.

    D. *If the fretting is over more than 50% of the fit surface, or, is part or all of the surface polished from contact with the shaft or housing?*

        Yes – The looseness is so severe that that the ring is rotating in the housing or on the shaft. With very low speed bearings the effect isn't serious but with typical bearings operating at more than about 300 rpm, heat transfer has been drastically reduced. As a result, the bearing's internal loads are increased and the life is reduced. In addition the movement results in vibration that increases the bearing load.

        No – There is movement and it is reducing the bearing life by increasing the loads and the operating temperatures, but it is probably not critical.

    Comment – As the fit worsens the bearing tends to rotate more in the housing. As the ring spins, it becomes polished and internal temperatures increase. Also the load isn't properly distributed on the ring and the possibility of a cracked ring increases (Photos 10.10 and 10.11).

    E. *Is there a black irregular surface on the housing or shaft fit, showing that the fit was extremely loose and the surface was heavily loaded?*

    Comment – If the fit is very poor and heavily loaded the surface will frequently spall and show surface fatigue damage from the heavy

PHOTO 10.10   This is a pair of angular contact ball bearings that were installed in a reducer. The highly polished surfaces show that they were both relatively loose and turning in the housing and we can see that the more polished lower bearing was looser.

PHOTO 10.11   The outer ring of a pump bearing that is so highly polished that reflections of the surroundings be seen in this mirror-like surface. Also, an inspection shows the upper edge of the ring is turning blue from the heat generated within the bearing and has been heated to about 250°C (≈500°F). The brown rings around the center of the ring are from fretting.

contact forces. The color of the shaft may vary some but the blackened shaft shown in Photo 10.12 shows the area had high humidity.

Sometimes this surface fatigue damage will include regular patterns such as that shown in Photo 10.13. People will assign all sorts of causes to this but it is only the repetition of the grinding pattern used in manufacturing superimposed on the loose fit. (This bearing came out of a truck and a dimensional check showed it was likely that the original fits were not what the bearing company specified.) A second interesting point that can be seen in this photo is that the side of the ring is highly polished in a band about 6 mm (1/4″) wide, from rotating against the shaft shoulder.

PHOTO 10.12   This shows a shaft where the two bearings were extremely loose and the result was this black irregular damage. (This is typical of a corroded surface fatigue.)

PHOTO 10.13   A tapered roller bearing with an interesting fretting pattern on the inner ring. The pattern resulted from variations in the grinding of the shaft contact surface.

Before leaving the subject of shaft and housing fits, ring cracking has to be reviewed.

One of the reasons a close fit is needed on the outer bearing ring is to distribute the fatigue forces over a larger area. If the ring is loose, the operating forces are concentrated into a smaller area and the possibility of cracking increases. (The poor fit increases the bending stress in the ring.) Photo 10.14 shows the outer ring of a self-aligning pillow block used on a fan and has a star-shaped crack toward the left side of the photo that started on the OD. The inspection found fretting on both the top and the bottom of the ring, showing the pillow block bore was not round and that there was considerable movement between the bearing and the housing. In addition, the black discoloration shows the bearing has been wet. (This puzzled the plant personnel until they realized that the bearing was occasionally subjected to humid outside air while the housing was still around 15°C [~60°F].)

PHOTO 10.14   The star fracture on the left side of the photo and the fretting, both symptoms of looseness, are readily visible on this ball bearing.

Similar fractures, because of multiple stresses, can happen on the inner ring and Photo 10.15 shows the inner ring of a spherical roller bearing. Looking at it we see:

1. There is a crack across the ring.
2. The shaft mounting surface is heavily fretted and there are polished circumferential bands that show it has been spinning on the shaft.
3. Like the bearing in Photo 10.13, the side of the ring is highly polished, almost mirror-like, from movement and contact against the shaft shoulder.

PHOTO 10.15   This shows the cracked inner ring of a spherical roller bearing. The polishing on the side and the smeared inner mounting surface prove the bearing has been loose and spinning on the shaft.

PHOTO 10.16   Both of these inner ring fits look suspicious, but the clear bands show the bearings were not rotating.

One last comment on the subject of shaft and housing fits is that occasionally the initial appearance will be deceiving. In Photo 10.16, the fits are not great and, looking at the fretting on the inner ring surface of the two bearings, the suspicion is that they are very loose. However, the clean unmarked area that was over a lock ring slot shows that, although there was some motion, they were not rotating. (These bearings had horrible outer ring fits and badly contaminated lubricants. The inner ring fit problem was a minimal contribution to their failure.)

  F. *Are the seals and/or shields in good condition? If not, why not?*
     Comment – Seals and shields are designed to keep contamination out of the bearing, and we know that even a small amount of water or dirt will reduce the bearing life because of increased surface fatigue stresses. Generally, lip seal life is only a fraction of the design life of a machine but, even after they degrade and lose contact, the seal lips tend to protect against water and larger particles. Inspect them and see if they were properly installed. (Were they installed to prevent contamination from getting in or were they supposed to prevent oil leakage out?)
     The best tool for seal inspection is probably a binocular microscope. Like the seal in Photo 10.17, look closely at the contact surface, searching for irregularities.

3. Next, look at the sides of both rings exteriors and answer the following questions.
   A. *Does the original grinding pattern show clearly around the entire surface?*
   Yes – The ring may have been moving against its restraint but the contact was not substantial.
   No – There has been movement and contact between the ring and the restraint.

PHOTO 10.17 This shows a view of a leaking seal lip at about 10X magnification. The equipment manufacturer had been complaining about poor seal quality but the circled area in the middle of the photo shows the seal lip was cut during installation at their plant.

B. *Are the sides of the rings polished? (See* Photos 10.13 and 10.15.)

Yes – The ring had been moving against the restraint and the contact was substantial.

No – There has been no movement or no contact.

C. *If there is polishing, is the pattern consistent with the mating piece pattern?*

Yes – That confirms looseness and relative movement.

No – Differences in the patterns will show some unique motion is present and require detailed further investigation.

D. *Are there any black spots or stains on the sides of the rings?*

Yes – There has been free water present.

No – Water has not contaminated the bearing.

4. Remove the seals and/or shields and visually examine the lubricant.

A. *Do the remains look discolored, contaminated, or oxidized?*

Yes – This shows the lubricant was not in good condition when the bearing failed.

No – The lubricant did not play a substantial part in the failure.

Comments – Analyzing the lubricant removed from a failed bearing is frequently challenging because the presence of deteriorated lubricant has to be weighed against the condition of the bearing at failure and one of the first questions that has to be asked is "Which came first, the deteriorated lubricant or the bearing failure?" It is extremely difficult to diagnose problems with bearings that have already had major damage such as that shown in Photo 10.18. This bearing shows some of the classical symptoms of an "inadequate lubrication" failure but how do we know that it didn't have other problems such as a poor fit, or brinelling, or improper metallurgy, or …

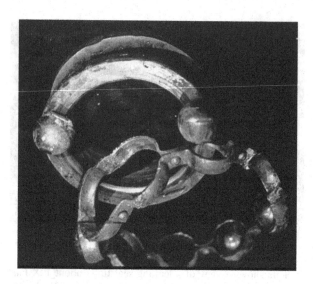

PHOTO 10.18   An absolutely destroyed bearing and it's not worth trying to do a failure analysis because of the severity of the damage can hide the underlying defects.

There will always be an argument between those persons who want to run a piece of equipment until the last possible moment before a tremendous catastrophic failure and those who want to shut the machine down and remove the part at the first indication of a serious defect. The first group is trying to maximize their immediate output while the second group, and I am one of them, is trying to learn from their mistakes. If we don't look at our errors and try to avoid them, what sort of progress are we going to make on improving reliability? Trying to do a failure analysis on a bearing such as that shown in Photo 10.18 is largely a waste of time!

    B.   Is there evidence of water?

        Comment – A crackle test will show only gross contamination but is relatively easy to perform. To perform the test, take a small sample of lubricant, a couple of drops of oil or a small dab of grease, and put it on a hot plate that is at 160°C +/– 5° (320°F +/– 10°F). If the sample "crackles" immediately after it hits the hot surface, it shows there is water in the sample. (The crackling is caused by the water boiling.) This test is only accurate down to about 0.1% water and we know that quantities below that can cause shortened bearing life, but it does give a good field guide.

    C.   Smell the lubricant. Has it been burned? Is there chemical contamination? Does it have a burned chemical smell or does it smell like the new grease that just came out of a grease gun? There are times when the lubricant will appear dry or powdered and this requires further examination. If the lubricant has been contaminated by dust or rust particles, these particles will frequently absorb any available oil. The appearance is that of an underlubricated bearing but many times the problem is contamination.

D. Take a small sample of the lubricant for a laboratory analysis.

Comment – Frequently a failed bearing will be blamed on "inadequate lubrication" or "lack of lubrication" when in reality there was plenty of lubricant before the bearing got hot. After the failure there are times when all that is left is the residue from burned grease and a careful inspection is needed to find the remains of any lubricant. On the other hand, there have been many times when a nervous oiler has added grease after the bearing has begun to fail. Be very careful during this phase of the investigation. If the bearing has failed with substantial damage and the grease looks like new, it is a good bet that it is and grease was added as or after the unit was shut down.

5. Look at the application to determine the loads and understand what the ball and roller paths should look like, then carefully separate the rings so those paths can be inspected. Place the fixed ring on a "lazy Susan" and slowly rotate it while watching the ball/roller path.

A. Is the ball/roller path where it was expected to be? If not, reanalyze the application to ensure your expectations were correct. Figure 10.5 shows some typical normal ball paths.

Ball and roller paths and load zones should be relatively uniform and variations indicate the effects of loads. Some of the key points to understand include:

- If they wander from side to side and vary in width, other than that expected from the normal loads, there is misalignment, ovality, or dimensional errors in the application.
- Probably the first thing that should be done is to carefully measure the width of the ball or roller path, then measure the rolling element width or diameter.
- There will be some variation with respect to load. (The heavier the ball is loaded the more it is deformed, and the wider the path.) In a lightly loaded application, the ball path should be about 0.2 times the ball diameter.

In a roller bearing the contact path should be the same width as the roller. Wider contact paths indicate dimensional variations.

One of the frustrations in writing this book is that I can't show you a video of a bearing inspection using a lazy Susan, one of best diagnostic tools imaginable. For years we struggled with trying to inspect and interpret the ball and roller path variations. Now we just put the bearing on the lazy Susan and spin it while varying the lighting and the size and variations in the contact paths, the trings, tell us what happened. Another approach is to roll the bearing across a table while looking at the paths, but the lazy Susan is a far better tool for accurate analysis – and it's inexpensive! (Figure 10.6)

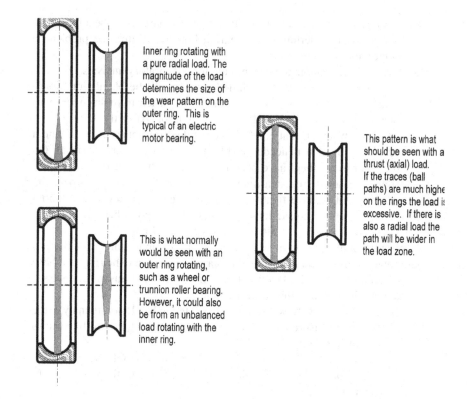

Inner ring rotating with a pure radial load. The magnitude of the load determines the size of the wear pattern on the outer ring. This is typical of an electric motor bearing.

This pattern is what should be seen with a thrust (axial) load. If the traces (ball paths) are much highe on the rings the load is excessive. If there is also a radial load the path will be wider in the load zone.

This is what normally would be seen with an outer ring rotating, such as a wheel or trunnion roller bearing. However, it could also be from an unbalanced load rotating with the inner ring.

FIGURE 10.5 A series of normal ball paths (traces) showing some typical ones to expect.

B. *Is there evidence of abnormal thrust loading, i.e., a path to one side of a ball bearing, polishing on the ribs of a tapered roller bearing, or unequal wear on the paths of a spherical roller bearing?*

Yes – There is an axial (thrust) load.

No – There is no significant axial (thrust) loading.

Two Comments –

1. In a deep groove ball bearing, the ball path may be off to one side and these bearings are designed to carry some thrust load. But, looking at a cross-section of the ring with the normal ball path at six o'clock, the path resulting from the thrust load should rarely be centered any higher than 7:30. If it is higher on the side of the ring the thrust load was likely excessive.

2. Analyzing the roller paths on spherical roller bearings is particularly important. The paths should appear essentially the same and uniform around the ring. These bearings are designed to take some thrust loading and there may be more wear on one of the paths, but their primary load should be radial and measurable wear on one path with little or no contact on the other is proof

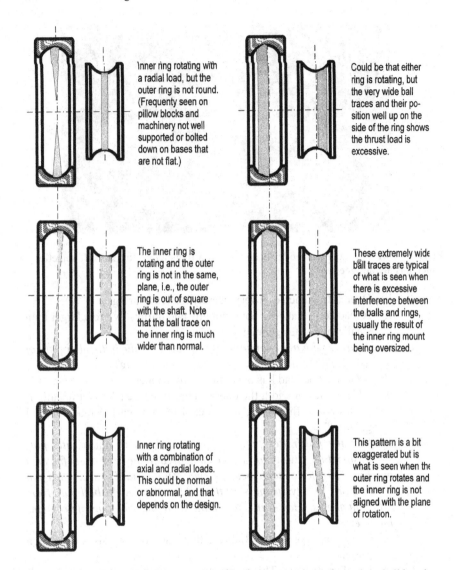

FIGURE 10.6 Six usually abnormal ball paths (traces). Although these show ball bearings, abnormal roller bearing paths are similar.

that the one path is carrying the entire load. Photo 10.19 shows the outer ring of a spherical roller bearing that was installed with no end float, i.e., the bearings were on a fan and both bearings were fixed so they couldn't move when the shaft expanded and contracted, and it is obvious from the photo that one ring carried the entire load.

   C. *In a bearing with a pure radial load, how large is the load zone?*
     i. Less than 45° indicates a lightly loaded bearing.

PHOTO 10.19 A spherical roller bearing with a tremendous thrust load, something for which it is not designed.

     ii. Over 120° indicates a very heavily loaded bearing.
    iii. A load zone that does not gradually and uniformly grow larger then smaller indicates geometry problems.

D. *In a bearing with a predominant thrust load – Is the ball/roller path of consistent width and location around the entire ring?*
    Yes – Good! The load has been uniformly applied.
    No – If it varies in width the question has to be asked, "Is it misalignment or is it the effect of expected loads from the machine and gravity?"

E. *Look at the ball/roller path. It should be smooth and lightly polished, sometimes even hard to see compared to the very finely finished machined surface of some bearing rings. (If it is hard to see, when you rotate it on the lazy Susan, shift the light to shadow the ball path from different angles.) However, a few of the common symptoms of problems include:*
    i. Black marks or spots – Indicate that water or corrosives have been present.
    ii. Highly polished – This mirror-like appearance is the result of skidding and may be caused by inadequate lubricant film and/or water contamination. Photo 10.20 shows a ball path that looks like a mirror and a careful inspection shows my digital camera. This was a bearing that was started without lubricant. It was greased after about three hours but the damage was already done. (The wide ball path is a symptom of other problems.)
    iii. Matte surfaced – This damage can result from fine contaminants and debris in the lubricant, from grossly excessive loads, and from the early stage of an electrical discharge failure. (The most common we've seen is the electrical discharge damage.) If it appears as a wide path essentially all the way around the bearing, it's most

PHOTO 10.20   This shows an extremely shiny and polished ball path, the result of skidding. Look closely at the center of the path, in line with the horizontal mark on the right side of the ring, and some scarring is visible.

likely excessive loads. Photo 10.21 is of a heavily loaded and badly contaminated ball path. Both electrical discharge and contamination tend to be centered in the load zone.

PHOTO 10.21   This matte and wide ball path that ran all the way around the outer ring is proof that this bearing was running with inadequate internal clearance, i.e., the shaft fit was too tight.

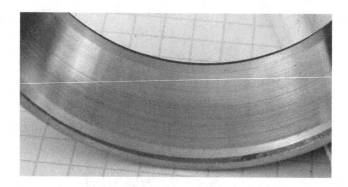

PHOTO 10.22 This roller path looks like it has been made with a heavy wire brush and that's because there was contamination (hard sand) in the lubricant.

    iv. Bruised – When hard foreign particles, such as debris from spalling, contaminate the surfaces and are rolled into the ball and ring surfaces, they leave indentations. These indentations are called "bruises".

    v. (If the damage happens during the manufacturing process before the bearings are finish ground, surface imperfections are called "dents".) The dings and damage in the upper roller path of Photo 10.19 are bruises.

    vi. Circumferential lines – Evidence of grit particles dragged along the rolling element path as shown in Photo 10.22

    vii. Spalled – A spalled surface is one where *fatigue cracks* have developed, and pieces of the ring have flaked loose. Below we list and discuss several of the more common spalling appearances. But before going on to them a quick look at the spalls in Photo 10.23 is in order. This photo shows two spalls and a ball path that is well centered, heavily loaded, and both bruised and highly contaminated. We can tell from the shape of these spalls that they are not just the result of Hertzian fatigue and we suspect other causes are present.

       – *Single small deep spall* – Typically the result of Hertzian fatigue, arcing from high amperage electrical current passage, or a single external blow. Photo 10.24 shows a Hertzian fatigue failure of a roller bearing. The spall began below the surface on the far right and grew to the left as pieces flaked out of the surface. Photo 10.25 also shows a Hertzian fatigue spall but this one is in the outer ring of a deep groove ball bearing. For both of these bearings a major clue that they are from Hertzian (end of life) fatigue is the relatively deep nature of the spalls.

         • If there are *multiple spalls* with same spacing as rolling elements the cause is almost always either brinelling or false brinelling, but there will be an occasional failure caused by electric arcing where the current flow is high enough to simultaneously arc through at several elements. Three photos showing this uniform spacing are below.

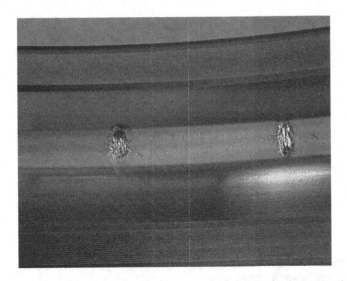

PHOTO 10.23 Two narrow spalls in a ball path that looks etched from the debris being rolled along it. (This appearance would make me suspect false or true brinelling or electrical damage, but further analysis would be needed.)

PHOTO 10.24 A deep and irregular Hertzian fatigue spall.

- The first, Photo 10.26, is a forklift truck spindle where the mechanic has misaligned and driven the bearing onto the shaft. When a bearing is *brinelled* the impact of the rolling elements being driven into the ring results in a series of dents that rapidly lead to multiple spalls.
- The second, Photo 10.27, is a view of a bearing with false brinelling. Brinelling is a denting process that happens in a single blow. *False brinelling* happens over time and is actually a welding and corrosion process that removes material

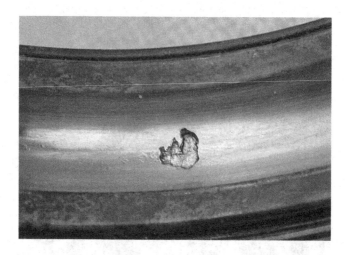

PHOTO 10.25 A single Hertzian fatigue spall. Metallurgical analysis may be needed to understand exactly why it started here.

PHOTO 10.26 A forklift truck spindle where either the bearing was dropped onto it or the mechanic beat the bearing into position. (Notice how uniform the marks are on the leading edge of the spindle where the rollers contacted it.).

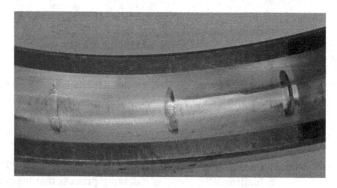

PHOTO 10.27 Brinelling happens with a single blow. False brinelling is the result of tiny vibrations over a long period of time.

from both the ring and the elements. In this example the outer ring has multiple depressions from *false brinelling* and if it had continued to run it would have failed rapidly.

- A very common location for false brinelling is where there are a series of machines that are on the same foundation or piped together and only one of them is run. The other machine(s) feel the vibration from the running ones and eventually false brinelling damages the bearings. (This can be eliminated by just rotating those spares every few weeks. Even a single rotation of the shaft is usually enough to prevent damage.)
- Photo 10.28 is of a pump bearing that shows advanced damage. Once a bearing has run for a while it is almost impossible to tell the difference between *brinelling* and *false brinelling*. (To get this example, we accidentally dropped a 500 kg [~1100 lb] pump while rigging it into position. It fell about six feet and landed on the shaft, and we were worried that we had brinelled the bearing. We didn't have a spare and put it in run, but vibration analysis on the unit immediately indicated that there was a problem. The pump ran for about nine months before it was removed from service on a planned outage.) The photo shows the regularly spaced damage that started with the brinell marks but has been multiplied by water contaminating the lubricant, resulting in many shallow spalls.
- *Continuous fine shallow spalled path* – The result of an inadequate lubricant film, either the load is too great, or the lubricant didn't do its job. This is also sometimes called "peeling" and is actually surface fatigue, where the spall starts on the ringway surface and grows inward.
- Photo 10.29 shows one path on a double row spherical roller bearing and the fine peeling is typical of a bearing with a

PHOTO 10.28 If we didn't know what had happened, it would be essentially impossible to tell if this damage started from true or false brinelling.

PHOTO 10.29   An excellent example of surface fatigue where the cracking starts on the surface and grows inward.

serious lubrication problem. In this example the lubricant was contaminated and couldn't do the job.
- Photo 10.30 is actually the same bearing as seen in Photo 10.20 where the excessive end thrust overpowered the lubricant.
- Photo 10.31 is a good comparison of the differences between the shallow *surface fatigue* on the left and deeper *Hertzian* fatigue on the right.
  - *Irregular "patches" of spalling* – A combination of forces. In the example Photo 10.32, it is a combination of false brinelling (or brinelling) and lubricant breakdown. It could also have been one of them combined with shaft or bore eccentricity. (Also notice that there was no load on the other ball path.)

PHOTO 10.30   Another example of surface fatigue.

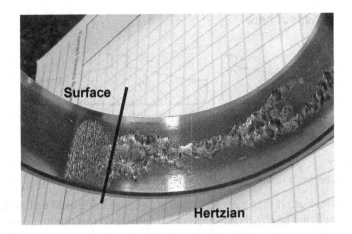

PHOTO 10.31   An example with Hertzian fatigue on the right and surface fatigue on the left.

PHOTO 10.32   A real challenge and the diagnosis would have to depend greatly on the location and operating practices. The thrust loading is obvious. The surface fatigue spalls may have started from true or false brinelling. Further analysis is needed.

F. *Carefully look at the color of the ring and the ball path. Is the ring the same color as when it was new? (Using the reflection of a fluorescent light, tilt the ring and watch the reflection travel across the rolling surfaces. Is there any tint or change in surface color?)*
   – A pale gold tint indicates the rolling path has been over 204°C (400°F).
   – A blue color indicates the ring has been well over 288°C (550°F).
G. *With roller, spherical roller, and tapered roller bearings, inspect the condition of the thrust shoulders. (Figure 10.7 shows the critical areas*

Inspect here

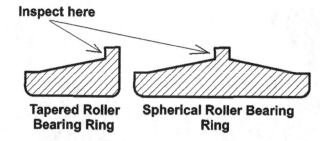

**Tapered Roller   Spherical Roller Bearing
Bearing Ring   Ring**

FIGURE 10.7 A sketch showing the critical location for inspection of roller bearings to determine if the thrust loading was more than the lubricant could withstand.

to look at.) If they have seen contact, the surfaces should be lightly polished, not heavily burnished or mirror finished from sliding contact.

APPEARANCE OF ELECTRICAL DAMAGE MECHANISMS

Before we go on to Step 6, we should discuss the effect of electrical currents on the ball path appearance. A common occurrence with low amperage, high voltage current leakage across the element path is *fluting*. Photo 10.33 shows an example of fluting in a 150 kW (200 hp) 1800 rpm motor bearing. The individual lines are caused by electrical arcing and remelting of the bearing steel.

A close up of some fluting damage can be seen in the magnified view of Photo 10.34. This is from a spherical roller bearing on a reducer and one can see that the flutes are actually areas where the surface has been removed. One additional interesting point in this bearing are the corrosion pits in the upper half of the photo.

There are times when *mechanical fluting* is confused with electrical fluting. Mechanical fluting results from periodic or constant pulsing loads, such as the load delivered to a bearing from the pounding of a poorly meshing spur gear. The bearing surface sees this persistent hammering and if the hammering always happens at the same positions around the ring, the surface eventually can be plastically deformed. Photo 10.35 shows an example of mechanical fluting removed from a large reducer. The faint dark lines across the roller path are the flutes.

PHOTO 10.33 A bearing from a motor controlled by a VFD showing the later stages of *fluting damage*. The fact that this ball path is extremely wide states there were other problems.

PHOTO 10.34 Fluting damage as seen through the microscope at low magnification. At high magnification the individual arc pits are visible.

PHOTO 10.35 Mechanical fluting is from repeated blows to the bearing. These faint lines are actually plastic deformation of the bearing ring.

Without metallurgical analysis, differentiating between the two types of fluting is difficult. Electrical fluting is generally better defined and there are many more flutes, but a metallurgical analysis can give a definite answer. If electrical fluting is present the damage will show as a white remelted area.

Photo 10.36 shows second type of electrical damage that is usually seen as a result of improper grounding during welding. The typical situation involves a welder not grounding close to the weld and a second ground path allowing flow through the bearing. The crew was welding on a pipe connected to a running pump and the resultant arced pits show each pulse of the AC welding current.

(Two notes on this bearing:

1. This type of damage can also occur during other high voltage, high amperage electrical abnormalities.
2. This bearing also shows contamination and corrosion damage.)

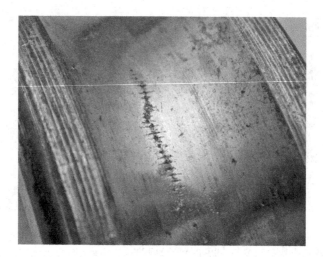

PHOTO 10.36 This is arcing damage from welding currents going through the bearing do to an improper ground, and the individual arc strikes show it's an AC welder. (To avoid this, it is critical that the welder be grounded close to the site of the welding. Using "building grounds" tends to cause these failures.)

If the welding shown in Photo 10.36 had been done while the machine wasn't running it is likely that the elements would have been welded to the rings and we have seen several examples of that.

Photo 10.37 is of a bearing ball that has been subjected to high currents. Like the fluting damage, positive proof of this can be found through a metallurgical examination.

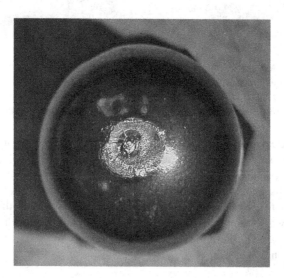

PHOTO 10.37 There is a temptation to say this happened because of a metallurgical defect, but it's actually the result of welding current passing through the ball when the machine wasn't running.

One expensive example of welding current being transmitted through the rolling elements occurred on a large crane in a wood yard. The plant scheduled some repair welding on the crane boom and the welders hooked their ground clamp to the base of the crane while pulling the electrode 100 feet in the air to the boom. They completed their welding and the operator mentioned that it was tough to get the crane to swing the first time after they were done. Several weeks later he mentioned that the crane swing was rough and "grumbling". An inspection of the 2 kg (4.4 pounds) bearing balls found that about half of them had suffered arcing damage from the welding.

6. Continuing with the physical inspection, place the rotating ring on a lazy Susan and slowly rotate it while watching the ball/roller path.

   Ask the same questions that were asked in Step 5.

7. Hardness test the rings and rolling elements.

   Comment – All of the bearing manufacturers we know of state that the rolling element and ring hardness should be between HRC 58 and HRC 62. Many of them also say the rolling elements should be a point or two harder than the rings but that isn't always true. If the pieces are softer than HRC 58 the reduction in hardness is a time/temperature reaction that can frequently be used to understand more about the failure progression. (Bearing steels start to lose their hardness when they exceed about 220°C (~425°F) and the longer at temperature and hotter the temperature the more the steel softens.

8. Inspect the cage paying careful attention to the surfaces the elements contact.

   A. Has there been contact on both sides of the element pockets?

      Yes – This is a situation that requires a bit of thought.

      i. Most bearings are in applications where they don't have enough of a thrust load to have the elements in contact with both rings all the way through 360. In these applications the rolling elements in the load zone drive the cage, but as soon as they leave the load zone they are driven by the cage. So, there may be some light wear on both sides of the cage pockets but there shouldn't be any plastic deformation. (You should not be able to feel a ridge of material along the top of the element pocket.)

      ii. The same holds true in a reversing application. There will be contact with the cage, but the surfaces should be lubricated to the point where there is only slight wear and there should be no plastic deformation.

      iii. However, if there is substantial wear, plastic deformation of the edge of the cage, or a cage fracture, it indicates there are either substantial reversing loads or poor lubrication. (These reversing loads generally occur only with torsional vibration and are a common problem source.) In Photo 10.38 this angular contact ball bearing cage shows tremendous plastic deformation. There are two obvious

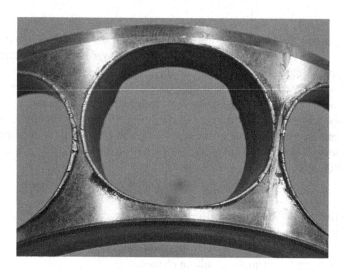

PHOTO 10.38   The plastic deformation of this battered cage proves it has seen grossly excessive loads and the lubricant hasn't been able to properly protect the cage.

  problems, the bearing was so heavily loaded the lubricant film broke down and there was a serious torsional vibration in the drive.
 iv. In a bearing with a combined load, i.e., a combination of radial and axial loads, there should be wear on only one side of the pockets. If there is wear on both sides it indicates there are definitely torsional vibration problems.
  No – This should be the case with most bearings.
B. In a roller bearing, is the cage wear uniform across the cage bars and around the entire cage?
  Yes – The geometry of the bearing is acceptable and the loads are consistent with good design.
  No – The bearing rollers are not rolling correctly. Either the lubricant film is inadequate or the geometry is incorrect. Photo 10.39 shows a spherical roller bearing cage. If the bearing had experienced pure

PHOTO 10.39   Another plastically deformed cage indicating either lubrication problems or excessive loads.

radial loads the cage might be slightly worn on the cage bars. But this cage shows heavy wear on both side of the pocket as well as wear on the near side of the pocket. This combination of symptoms shows there are serious lubrication problems and possible reversing loads during each rotation.

In one application the plant was having frequent failures of the roller bearings in a reducer driven by a variable speed steam turbine. The reducer was, in turn, driving a reciprocating compressor. Inspection of the bearing cages found the cage bars were failing from fatigue and several of them had broken out of the cage. Further analysis found that at some speeds there was a torsional resonance within the reducer and the forces were enough to break the cage bars. The solution was to limit the speeds the turbine could run at.

C. In a ball bearing, is the cage wear uniform around the entire cage?

Yes – The geometry of the bearing is acceptable and the loads are consistent with good design.

No – The balls are not rolling uniformly during the rotation and the bearing's operating geometry is incorrect.

9. Inspect the rolling elements.

For ball bearings

A. Is the ball surface uniformly polished with a fine ground surface?

Yes – Good. This is indicative of good operating conditions and satisfactory lubrication.

No – The common conditions that contribute to rolling element damage are:

- Black spots and/or blackened grooves – The black stains are corrosion products and indicate water has been present when the bearing wasn't rotating. The black smears on the balls in Photo 10.40 are proof that water has been present.

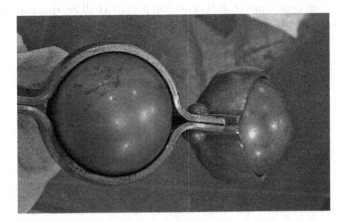

PHOTO 10.40 The black spots are corrosion caused by water in the bearing. (One of the major problems resulting from corrosion of hardened steels is hydrogen damage.)

- Highly polished mirror-like surface – The result of skidding and a poor lubricant film. Frequently evidence of water during operation.
- Small circular scratch marks – The result of hard, relatively large, contamination particles that were embedded in the cage.
- Matte, lightly sandblasted appearance – Three likely causes are
- Fine contamination in the lubricant
- High frequency, low amperage electrical discharges
- Extremely heavy loads breaking down the lubricant film

Comment – A lubricant wear particle analysis should be able to tell if the lubricant was contaminated. However, both of the first two causes will result in fine particle contamination. A simple metallurgical examination where the ring is cut and the cross-section polished and etched should reveal if there is any rehardening from electric arcing.

B. Are there obvious bands or stripes around the balls?

No – Good. The elements should roll freely and there should be no pattern of scratches or circular bands.

Yes – This could be the result of several conditions as follows:

- Fine circular scratches – These usually result from hard debris particles caught between the ball and the cage.
- Discolored circumferential bands – Typically the result of inadequate clearance between the inner and outer rings. The bands form when the balls can't freely rotate as shown in Photos 10.41 and 10.42.

For roller bearings

Comment – The contact surface of a roller is much larger than that of a ball and should be a straight line. As a result, roller bearings place more severe demands on both the lubricant film and the shaft and housing geometry. For example, a slightly tapered fit that would result in an imperceptible minor shift in the ball path of a ball bearing will result in significantly increased stresses on a roller bearing.

PHOTO 10.41 These three balls were pinched between the two rings so they couldn't rotate freely. This is most often the result of the inner ring fit being too large.

PHOTO 10.42 If the internal interference is even worse, the rolling elements frequently shatter like the ones shown in this photo. (The replaced motor shaft bearing fit was significantly oversized.)

A. Is the roller surface uniformly polished with a fine ground surface?

Yes – Good. This is indicative of good operating conditions, good alignment, and satisfactory lubrication.

No – The common conditions that contribute to rolling element damage are:

- Black spots or patches – The black stains are corrosion products and indicate water has been present when the bearing wasn't rotating.
- Highly polished mirror-like surface – The result of skidding and a poor lubricant film. Frequently evidence of water during operation.
- Circular scratch marks around the rollers – The result of hard, relatively large, contamination particles wedged into the cages.
- Matte, lightly sandblasted appearance – See ball bearings, earlier

B. Do the ends of the rollers have circular worn bands, stripes, or small semicircular marks?

No – Good. The ends should be as originally ground or very slightly polished. They should not have a mirror finish or have a pattern of scratches or bands.

Yes:

- Highly polished roller ends almost always result from either:
  - Improper (low) lubricant viscosity.
  - Moderately elevated thrust loads that have destroyed the lubricant film and resulted in metal-to-metal contact.

Comment – Highly polished roller ends with circular score marks or "semicolons" on them as shown in Photos 10.43 and 10.44, show that

PHOTO 10.43   This photo and the next show damage to the ends of the bearing rollers that shouldn't be happening. Either the load is excessive or they are using the wrong lubricant.

PHOTO 10.44   The swirl pattern on the ends of the bearing rollers shouldn't be there. Again, either the load is excessive or they are using the wrong lubricant.

either substantially elevated thrust loads or very poor lubrication has occurred. (Rollers are not made with pretty swirl patterns on them.)

ROLLER AND TAPERED ROLLER BEARING MOUNTING SURFACES

The accuracy requirements of roller bearing shaft and housing fits are much more demanding than those of ball bearings. If a ball bearing ring fit varies by 0.02 mm (0.0008″) across the width of the bearing the rolling path is shifted a miniscule amount toward one side and the ringway radius may vary a little. But in a roller bearing the design calls for the load to be distributed across the entire ring and that misalignment can have a serious effect on the fatigue stresses. The magnitude of the effect is a function of the size and load and larger units are less affected but any visible difference in the roller paths is cause for concern.

Photo 10.45 is of a roller bearing with the ring bore causing distortion and a heavier load on the side toward the reader. In this example there is also mechanical fluting with the individual flutes indicated by the black marks.

Photo 10.46 shows the roller paths from a spherical roller bearing that shows a pure radial load that was well divided between the paths and there was no measurable deformation of the outer ring. (Unfortunately, the discoloration tells us that the ringway has been very hot and the highly polished surfaces show the lubricant film was nonexistent.)

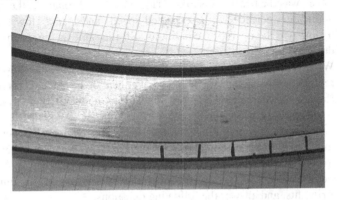

PHOTO 10.45   The lack of support has distorted the ring and the roller pattern runs toward the viewer. Also, the black lines are in line with the mechanical flutes.

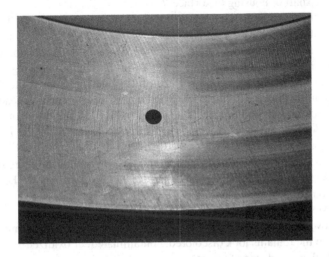

PHOTO 10.46   The pattern shows this spherical roller bearing has a very well divided radial load.

THE DETAILED ROLLING ELEMENT BEARING FAILURE ANALYSIS PROCEDURE

As a guide, the nine steps are printed below without the comments and illustrations.

1. Understand the background, i.e., the bearing's history.
   A. Look at the housing and the bearing's surroundings. Are they clean and uncontaminated?
   B. Measure the bearing housing bore and the shaft dimensions in at least three positions on each side of the contact. Compare these readings with the specifications
   C. What is the range of temperatures and relative humidity readings in the area? Is the shaft hotter than the bearing?
   D. When was the bearing installed? How fast does it rotate? Is the bearing application correct for the expected speeds and loads? How long has it run? Are the operating conditions still the same as the machine was designed for?
   E. What is the lubricant, what is the relubrication schedule, how is the relubrication accomplished, and what education have the oilers had?
   F. Review the maintenance records, including the vibration analysis and/ or oil analysis history on the machine
   G. Speak with the operating, maintenance, and engineering personnel involved with the machine. Ask for their ideas on what could have contributed to the problem.
2. Inspect the contact surfaces of the inner and outer rings, i.e., the inner and outer ring fits, and answer the following questions.
   A. Are there any black stains on the rings?
   B. Other than water stains, is there any visible fretting or discoloration of the shaft or housing fit surfaces?
   C. Is the original grinding pattern still visible over more than 50% of the surface?
   D. If the fretting is over more than 50% of the fit surface, is part or the entire surface polished from contact with the shaft or housing?
   E. Is there a black irregular surface on the housing or shaft fit, showing that the fit was extremely loose and the surface was heavily loaded?
   F. Are the seals and/or shields in good condition? If not, why not?
3. Next, look at the sides of both rings exteriors and answer the following questions.
   A. Does the original grinding pattern show clearly around the entire surface?
   B. Are the sides of the rings polished? If there is polishing, is the pattern consistent with the mating piece pattern?
   C. Are there any black spots or stains on the sides of the rings?
4. Remove the seals and/or shields and visually examine the lubricant.
   A. Do the remains look discolored, contaminated, or oxidized?
   B. Is there evidence of water?
   C. Smell the lubricant. Has it been burned? Is there chemical contamination?

D. Take a small sample of the lubricant for a laboratory analysis.

5. Look at the application to determine the loads and understand what the ball and roller paths should look like, then carefully separate the rings so those paths can be inspected. Place the rotating ring on a "lazy Susan" and slowly rotate it while watching the ball/roller path.

A. Is the ball/roller path where it was expected to be?

B. Is there evidence of abnormal thrust loading, i.e., a path to one side of a ball bearing, polishing on the ribs of a tapered roller bearing, or unequal wear on the paths of a spherical roller bearing?

C. In a bearing with a pure radial load, how large is the load zone?

D. In a bearing with a predominant thrust load – Is the ball/roller path of consistent width and location around the entire ring?

E. Look at the ball/roller path. It should be smooth and lightly polished, sometimes even hard to see compared to the very finely finished machined surface of some bearing rings. (If it is hard to see, when you rotate it on the lazy Susan, shift the light to shadow the ball path from different angles.) However, a few of the common problem symptoms are:
   - Black marks or spots – corrosion
   - Highly polished – skidding
   - Matte surfaced – skidding
   - Bruised – debris in bearing
   - Circumferential lines – debris in bearing
   - Spalled – sequence shown below
      - Single small deep spall – end of life? Hertzian?
      - Multiple spalls – from brinelling or false brinelling
      - Continuous fine shallow spalled – lubrication problems
      - Irregular "patches" of spalling – likely started with brinelling or false brinelling

F. Carefully look at the color of the ring and the ball path. Is the ring the same color as when it was new? (Using the reflection of a cool white fluorescent light, tilt the ring and watch the reflection travel across the rolling surfaces. Is there any tint or change in surface color?)

G. With roller, spherical roller, and tapered roller bearings, inspect the condition of the thrust shoulders.

6. Continuing with the physical inspection, place the fixed ring on the lazy Susan and slowly rotate it while watching the ball/roller path.

Ask the same questions that were asked in Step 5.

7. Hardness test the rings and rolling elements.

8. Inspect the cage paying careful attention to the surfaces the elements contact.

A. Has there been contact on both sides of the element pockets?

B. In a roller bearing, is the cage wear uniform across the cage bars and around the entire cage?

C. In a ball bearing, is the cage wear uniform around the entire cage?

9. Inspect the rolling elements.
   For ball bearings
   A. Is the ball surface uniformly polished with a fine ground surface?
   B. Are there obvious bands or stripes around the balls? – Restricted motion – Fit too tight or dirt?
   For roller bearings
   A. Is the roller surface uniformly polished with a fine ground surface?
       Are there obvious bands or stripes around the balls? – Fit too tight or dirt?

FINAL COMMENTS ON FAILURE ANALYSIS

As with everything else that fails, there are almost always multiple causes. Bearings are extremely complex devices with multiple components machined to extraordinary tolerances and one unfortunate result is that it is easy to accidentally overlook a major contributor to the failure. Sit down with another person and look at the nine points discussed earlier in the chapter. Carefully answering the questions will lead you to the solution.

One more amazing bearing to look at – Photo 10.47 is of a bearing that didn't have a catastrophic failure like most that are shown in the book. It is out of a huge 7500 kW (10,000 hp), 3600 rpm motor that was put on the sixth floor of a building. 7500 kW motors aren't small and they aren't light, and what do you think the building steel did when the motor was installed on it? And when that building steel deflected, what do you think happened to the roller bearings that were supporting the shaft? The misalignment pattern was impressive and they went through a set of bearings about every five or six years.

They were doing vibration monitoring and could shut the machine down before it was a catastrophic failure (the good thing was that they had a circulating oil system on the bearings), and what was amazing to me was the amount of wear on this inner

PHOTO 10.47 The 350 mm (~14″) ID inner ring from the shaft of a 3600 rpm motor. The rollers have worn a path over 0.5 mm (0.020″) deep into the ring and it's been a bit warm!

ring. The running trace has worn more than 1/2 mm (0.020″) off the surface of the ring and you can see that it's been a bit warm.

Ironic is that, if they had made the machine with self aligning plain bearings it would probably be on its first set a bearings, but that would have increased the initial cost.

BIBLIOGRAPHY

Harris, Tedric, *Rolling Bearing Analysis* (3rd edition), John Wiley & Sons, New York, 1991, ISBN: 0-471-51349-0.

Life Factors for Rolling Bearings (2nd edition), Edited by Erwin V. Zaretsky, Society of Tribologists and Lubrication Engineers, Park Ridge, IL, 1999.

Rolling Bearings, NSK Rolling Bearing Catalog No. E1101c, Tokyo, Japan, 1999.

Technical Report, NSK Motion and Control Catalog No. E728g, Tokyo, Japan, 2013.

11 Gears

The design and operation of gears is more complicated than almost any of the common mechanical devices. They incorporate cantilever fatigue loading of the teeth similar to a beam in bending, Hertzian fatigue loading of the contact surface similar to a rolling element bearing, plus the lubrication demands of a pair of surfaces that involves both sliding and rolling. They are incredibly complex and their failure analysis isn't for the faint of heart or for those whose primary athletic ability lies in jumping to conclusions. But, from the basic principles covered in the earlier chapters and careful inspection of the failed components, determining the source of most gear failures isn't terribly difficult. However, to help in their understanding, the first part of this chapter explains some of the more important details of gear design and function.

GEAR TERMINOLOGY

Gear operation involves both rolling and sliding contact. The first true gears were designed involving only rolling contact but that proved impossible to sustain in a real world environment. As a result, almost all the gears we see have *involute* profiles and, as they contact, they essentially slide into mesh, involving more rolling as the teeth approach the pitch circle where the contact motion is pure rolling, then as they go out of mesh the high proportion of sliding movement returns.

(An involute profile is the shape developed by a point on the surface of a circle as the circle rolls along a surface and this gives us the typical curved shape of a gear tooth. There are other tooth profiles but few of them are in wide industrial usage.)

There are dictionaries of gear terms and much of the terminology is truly specialized. But to anyone but gear specialists, designers and manufacturers, most of the terms generally aren't needed and the ones shown in Figure 11.1 are by far the most common.

The gear terms used in this book are:

- *Active profile* – This is the contact surface of the tooth. It includes the addendum plus most of the dedendum.
- *Addendum* – This is the top half of the tooth.
- *Backlash* – The effective clearance between the mating teeth.
- *Circular pitch* – The distance between identical points on adjacent teeth – as measured along the pitch diameter.
- *Dedendum* – The lower half of the tooth. It is essentially the same as the addendum except the root clearance is added to it.
- *Gear* – The gear, or the *bull gear,* is the driven unit. It is usually, but not always, the larger of the two gears.

Important Gear Tooth Terminology

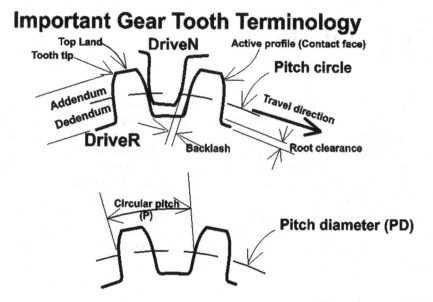

FIGURE 11.1 These two sketches show most of the common gear terminology.

- *Module* – Usually a metric term, this is a measurement of a gear tooth size and is the pitch diameter divided by the number of teeth
- *Pitch diameter (or pitch circle)* – This is the ideal line the meshing gears intersect along. It is the ideal diameter of the gear.
- *Pinion* – The driving gear that is usually, but not always, the smaller gear.
- *Root* – The "almost flat" section between two adjacent teeth.
- *Root clearance* – The clearance between the top land of the one unit and the root of the other is the root clearance.
- *Thermal power (or horsepower) rating* – Reducers are normally rated on the amount of power they can physically transmit with exceeding industry standards. The thermal rating is the amount of power the reducer can transmit without exceeding the operating temperature shows in the standard's guidelines.
- *Tip or top land* – The flat top of the gear tooth.

Another characteristic shown in Figure 11.2 is the *pressure angle*. This is the angle as measured from the tangent to the gear, on which the force is ideally transmitted. In the early 1900s, $14\frac{1}{2}^0$ gears were by far the most common. As time went on the teeth became shorter and fatter (and stronger in a bending mode) and 20^0 gears became the pressure angle of choice. Today many reducers use 25^0 teeth, and in specialized cases, even higher pitch angles are sometimes used because the teeth are stronger and less likely to fail catastrophically. The downside of these higher pitch angles is that the separating forces between the gears become higher and the gear train becomes slightly less efficient.

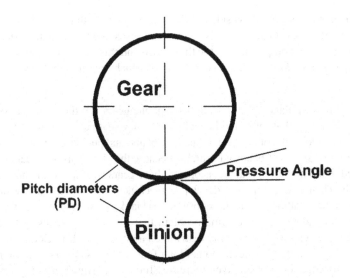

FIGURE 11.2 Shows the pressure angle of a pair of gears.

TYPES OF GEARS

Looking back in history, the first "gears" were wooden devices where the spokes off two hubs meshed with each other to transmit power and synchronize the equipment. As the industrial revolution began the idea of actual teeth working together was developed and these gears were straight tooth *spur gears* as shown in Figure 11.3.

One of the weaknesses of spur gears is that the power pulses are discrete, i.e., as each pair of teeth mesh a power pulse is delivered, and they are relatively noisy and rough. An early solution to this was the *helical gear*. With helical gears there is more than one tooth in mesh and the resultant power transmission is much quieter and smoother

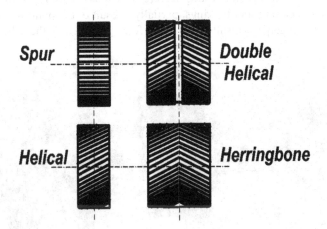

FIGURE 11.3 Four common gears. Although many gears today look like *herringbone* gears, they are actually double helical.

than with spur gears. The downsides of the helical gear are that there is a very slight decrease in efficiency and there is also a resultant thrust load. So, the next step in gear development was to compensate for the thrust load and this resulted in the development of *herringbone* and *double helical* gears, both of which are thrust balanced.

The first thrust-balanced gears were herringbone gears and they were cut on the same machinery that made the helical gears. They first cut one side of the gear blank, then took it out of position, rotated the blank, and put it back in the machine. The result was a thrust-balanced gear, but because the blank was rotated and repositioned, it did not have great precision and those gears were noisy and not ideally smooth. Then gear companies developed machinery that would cut the double helical pattern without having to reposition the blank and with great precision.

This was essentially the same problem that the bearing industry faced in that the stresses should be evenly distributed across the rolling elements, or in this application, the gear teeth, to get long fatigue life. As gears become more accurately machined, their lives increase. Today superfinishing of gears has enabled them to reliably transmit power loads that were inconceivable even thirty years ago.

A practical example of the difference between helical and spur gears can be seen in the typical standard shift car or truck. Driving the car in reverse produces a whining noise that doesn't happen in first gear. The gear ratios are almost identical but the helical first gear operation is smooth and quiet while the spur gear used for reverse produces a whining noise from the individual power impulses.

Photo 11.1 shows a "cluster gear" assembly from a manual automotive transmission. There are the four helical gears used for forward speeds while the one spur gear is reverse. (A second interesting feature of this assembly is the difference in the size of the tapered roller bearings, partially a result of the point where the load is delivered but largely the result of the thrust force produced by the helical gears.)

PHOTO 11.1 This is a "cluster gear" from a four-speed automotive transmission. The small spur gear is reverse.

PHOTOS 11.2 AND 11.3   A bevel gear on the left and a spiral bevel gear on the right.

Spur, herringbone, and double helical gears have to be used with parallel shafts. Also, most helical gears are used with parallel shafts but there are many other gears designed for use on non-parallel shafts. Two of these are shown in the photos while others are described in the text.

Photos 11.2 and 11.3 both show gears typically used in right angle drives. The bevel gear is essentially a right-angle spur gear and is similarly rough and noisy. The spiral bevel gear, with multiple teeth in mesh is similar to a helical gear in its smoothness.

Other common gear types include *hypoid gears*, which are variations on spiral bevel gears set up so their centerlines don't intersect, and *worm gears* which involve pure sliding motion, as opposed to the rolling and sliding motion of the other gear designs.

TOOTH ACTION

Before analyzing the stresses that occur within the gears, a review of the relative tooth movement and contact will make the wear and fatigue action much easier to understand. Three phases of this relatively complex contact are shown in Figure 11.4 and, as can be seen by the notes on the right side of the figure, the contact combines both rolling and sliding motions.

As the teeth first engage, the tip of the driven tooth slides down the driving tooth but at the same time the contact point is rolling upward. The combined motion yields a low relative speed at the contact point and, in many applications, there isn't enough velocity to create an adequate lubricant film. As movement continues, the contact point reaches the centerline, the tooth action is pure rolling in the upward direction and, with this increased speed, a thicker lubricant film is developed. As the contact continues further along, the rolling and sliding motions are in the same direction and a substantial lubricant film is developed.

However, as the contact point slides downward along the driving tooth surface, those contact areas that are in tension are subjected to fatigue damage. However, at the same time, the contact areas of the mating tooth are in compression and fatigue doesn't occur on them. Interestingly, later in the cycle, in the latter phases of contact, the roles are reversed and the driven gear dedendum can suffer fatigue damage. This combination of motions, with its effect on the lubricant film, plus the fatigue stresses, i.e., the negative and positive sliding stresses, explains why damage such as pitting and surface wear is most often seen in the lower halves the driving and driven teeth.

Tooth Position and Contact Action

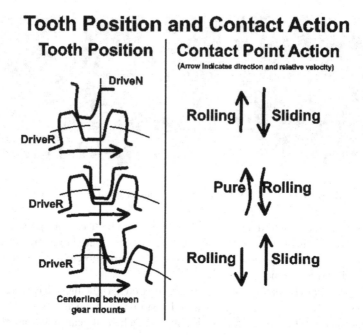

FIGURE 11.4   It's important to realize that the teeth initially slide into contact, then there is a changing combination of rolling and sliding that becomes pure rolling at the intersection of the pitch lines. Then the reverse happens as the teeth go out of mesh.

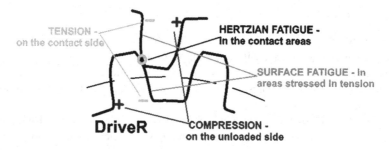

FIGURE 11.5   The stresses on gear teeth.

Not only are the teeth subjected to the Hertzian fatigue loads, they also see a cantilever bending load and the combined loads are shown in Figure 11.5.

In an interesting verification of the idea that fatigue cracks cannot occur in compression, we once tracked a pair of large cracks on the back side of a heavily loaded cast steel through hardened gear. The cracks were the result of manufacturing errors and did not have measurable growth over the years that we watched them while the contact side of the teeth had substantial wear.

LOAD AND STRESS FLUCTUATIONS

Theoretically the power transmitted by the driving teeth should be a constant at some level, whether it is 100 hp, or 500 kW, or 1 kW, and the teeth should all be loaded identically. But what actually happens during operation is that the forces fluctuate during each revolution. Some of the sources of these changes are:

- Coupling misalignment
- Gear misalignment
- Input power variations
- Gear and pinion eccentricities
- Machining errors
- Torsional vibration and resonances

The result of this is that a graph of the load, instead of being a flat straight line, actually varies as shown in Figure 11.6. In this example there are two gears with the same average load, but the peak load of one is 21 units while the other is 26 units. The one with the 26-unit load will likely last about half as long as the other one, because it's the peak tooth load that does the major damage.

Of special concern are those applications where damage (wear or polishing) is seen on the "non-contact" side of gear teeth. The damage could be the result of misalignment (which will be dealt with later) but more serious is that it could indicate there have actually been reversing loads and tremendous stresses on the teeth.

Photo 11.4 shows a 10″ diameter cam drive gear from a 3000 kW (4000 hp) gas engine with the teeth worn on both sides. The gear rotates clockwise and is supposed to drive in only one direction but this wear is positive proof of the reversing loads that are occurring within the gear train. Looking at both the gear and the values in Figure 11.7, it can be seen that the peak stresses resulting from reversing loads *must be more than twice* the average driving load and that serious damage can be occurring throughout the gear train. (In motor-driven gear sets we have found high resolution, i.e., capable of recording multiple times per second, data recorders and power meters to be invaluable diagnostic tools.)

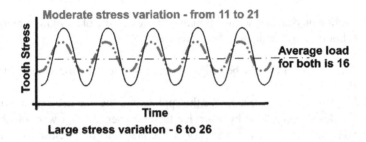

FIGURE 11.6 This shows a graph of the actual tooth stress for two gears, both driving a 16 hp or kW load, but with very different peak loads on the teeth, and it the peak stress that causes the major wear.

PHOTO 11.4   This gear drives in one direction, but there is wear on both sides of the teeth. That tells us that the teeth are seeing loads more than twice what the gear was designed for.

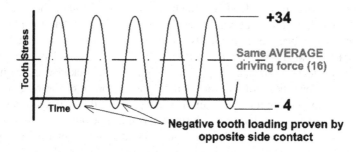

FIGURE 11.7   A graph of the tooth loading on a gear such as that shown in Photo 11.4. The contact on the inactive profile shows the gear must be seeing reversing forces and the peak stresses must be more than twice the design stress.

In the section on bearings we mention that the design life of an AGMA reducer is 5000 hours. Two comments about this are:

1. The average life of the bearings is usually shorter than the average life of the gears.
2. To properly size a reducer for the application, the practice is to multiply the design drive load by a service factor. When we first were involved with industrial reducer the service factors were often close to 3, frequently from 2.6 to 2.8. As a result of competitive pressures, the current

practice is to use service factors much lower than that, and we've seen them rated as low as 1.3. The reason this is important goes back to the effect of stress on the Hertzian fatigue loads in the bearings and the fact that cutting a bearing load in half increases the life by a factor of from 8 to 10. So, if a service factor of 2 is applied to a reducer that has all roller bearings and was rated for 5000 hours, the new bearing L_{10} life will be 50,000 hours (~6 years) assuming good lubrication and operating conditions. However, if there are 10 bearings in the machine it also effectively guarantees that there will be a failure in less than five years. In my opinion, increasing the service factor to close to 3 is a very inexpensive up-front investment in improved reliability.

GEAR MATERIALS

The first spur gears were made from wood and there are still some low speed and lightly loaded gears around that have maple teeth bolted into cast iron hubs. However, they are not common and most of the larger gears are metal but polymers have made substantial inroads and an interesting example of a blend of old and new technologies is the use of nylon teeth and steel hubs to quietly replace the cast iron gears driving the dryers on older paper machines. Of the all-metal gears, most are steel although there are many cast iron and bronze units in operation. Cast iron gears are mostly used on older machinery where there are relatively light tooth loads, but they are also used on some specialized equipment such as large rotary vessels and filters. The largest industrial use of bronze gears appears to be on worm gear applications where hardened steel worms are commonly mated against bronze worm wheels.

Over the years there have been changes in the metallurgy of steel gears and, as the metallurgy changes, the deterioration and failure appearances change. Therefore, in the failure analysis of steel gears, it is important to understand both the hardness and the metallurgy of the teeth.

The gears originally used in most industrial applications were *through hardened* and were designed to wear with age but by the 1930s *surface hardened* teeth began to appear. The terms can be confusing and defining them:

- *Through hardened* is an industry term that means the hardness of the gear tooth is consistent throughout the tooth, even though the steel may not have been hardened at all. Most "though hardened" teeth are relatively soft and almost all are less than HRC 40. (Also:
  - The hardness may not be exactly consistent throughout the tooth, but the difference is trivial, i.e., the interior may be a couple of HRC points lower because it takes longer to cool during heat treating.
  - Most polymer gear failures can be analyzed using the same techniques as those used on through hardened metal gears.)

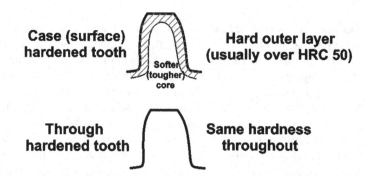

FIGURE 11.8 Many years ago gear teeth were through hardened, or not hardened at all. Today almost everything, except for very large gears, uses case hardening and it's very important to recognize the difference.

- As shown in Figure 11.8, *case* or *surface hardened* gears have a very hard and wear resistant outer surface with a much softer core. The surface hardness is almost always well above HRC 50 and they are designed to show essentially no wear at all, even after years of operation.

In analyzing a worn or failed gear, the place to start is to understand the gear hardness, but this is sometimes difficult in a field application. In those cases we frequently check the tooth surface with a file. If the file skids, the steel has to be hardened and any surface damage more serious than mild micropitting is cause for concern.

The "cluster gear" shown in Photo 11.1 was the source of a very practical gear metallurgy lesson for me. It happened in one of our cars shortly after the car was out of warranty and, as you can understand, I was more than a little frustrated. It pulled the gear out of the transmission and took it to a friend of mine who was the chief metallurgist at a local gear manufacturer and the second I showed it to him I knew he was also frustrated. His company had made the gear and their analysis found:

- This was a heat treated steel gear.
- In the heat treating process, the gear is heated to about 870°C (1600°F) and then quenched. It then is immediately put into a tempering oven for about an hour to reduce the stress between the case and the core.
- In the production of our cluster gear, the person responsible didn't put the gears into the tempering operation for a long enough time that microcracking began between the case and the core.
- Our normal operation of the car added stress that, with the locked in residual stress and the stress concentrations from the microcracking, caused a premature failure of the gear.

> The education from that lesson has paid off several times. One of them is shown in Photo 11.20 where the manufacturer instantly agreed with our request for a no-charge replacement.

In the early industrial cast iron gear applications, the gears were frequently run without grease or oil and the graphite in the iron served as a lubricant. (That is still done with many lightly loaded cast iron gears.) As time went on designers realized that harder materials wear more slowly. However, the trend toward harder materials was slowed by the facts that not only does impact strength decrease as the hardness increases, but also harder gears require closer tolerances in manufacturing and operation to prevent breakage. However, a big advantage of surface hardened gears is that they are smaller and lighter, and therefore less expensive, and can convey the same power in a much smaller package.

Most of the surface hardened gears in industry are used in enclosed reducers. In North America almost all of the reducers made before 1960 used through hardened gears (that really were not very hard) but some of the European gear manufacturers began to build reducers with case hardened gears shortly after World War II. By 1970 it was common to see American manufacturers making reducers with surface hardened gears for use up to about 75 kW (100 hp). By 2000 most 750 kW (1000 hp) reducers had case hardened gears and I've recently seen reducers rated at over 1500 kW (2000 hp) with all case hardened teeth. (We should add that, in the 1960s and 1970s, there were some very successful North American manufacturers that made large reducers with case hardened pinions and through hardened gears.)

In looking at the capabilities of the different gear metallurgies, note that gear ratings are very dependent on geometric conformity and surface finish. Going to higher quality machining, with greater precision and smoother surface finishes allows more power to be transmitted within the same physical configuration. With this in mind, the performance differences between the two basic metallurgical structures are:

Surface Hardened Gears
- Provide more power in the same size package
- Are used on almost all mobile equipment
- Demand greater precision in manufacturing
- Much less tolerant of surface damage, impact loads, and weak support structures.

Through Hardened Gears
- Used on large open gear sets and reducers where great precision is difficult
- Are much more tolerant of shock and impact loading
- Can tolerate substantial wear before failure

In addition, because they maintain their original geometry, case hardened gears usually run quieter and slightly more efficiently than their worn through hardened counterparts.

TOOTH CONTACT PATTERNS

In the discussion above the need for good alignment was mentioned. Looking further at the stresses between two mating teeth, Figure 11.9 shows the ideal stress on a gear tooth and what actually happens as the contact pattern changes.

This shows the importance of good tooth contact but in understanding it several points that should be mentioned are:

1. We will assume the contact in the first panel of Figure 11.9 is as good as it looks and plot the stress as being uniform across the tooth.
2. In the second panel there is linear contact all the way across the tooth, but it becomes weaker and just ends at the end of the tooth. In this example, in order for the driving force to be the same as the first panel, the maximum contact stress on the left side has to be twice the ideal (average) stress.
3. The third panel has the contact ending at mid-tooth. From this, the peak stress has to be at least four times that of an ideal tooth.

In analyzing gear tooth contact patterns one serious caution is that, because of elastic deformation, the contact stress is always more severe than it appears to the eye. For instance, if the teeth were operating like they appear in the center panel of Figure 11.9 and were inspected during a no load operation, they might look like those in the panel on the far right. This happens because the support structure and the teeth deflect and the contact looks much less damaging than it actually is.

An example of this is shown in Photo 11.5 and this was our first recognition of *"dynamic soft foot"*. The photo shows a gear from a 300 kW (400 hp) reducer that was mounted on a pair of 760 mm (30″) I-beams. The reducer was brand new and it looked perfect, but after running for six weeks or so, a check found there was heavy pitting on one side of the gears. Our initial thought was that

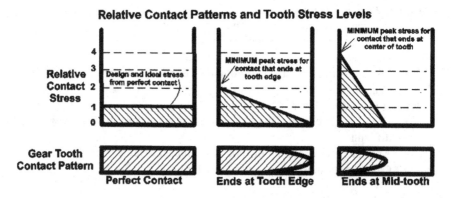

FIGURE 11.9 This shows the relative tooth stress for three alignment conditions, but is actually an idealized diagram. In reality, because of elastic deflection, the actual contact stress can be much higher.

PHOTO 11.5 This is an intermediate pinion and, after six weeks of operation, the pitting shows the severity of the misalignment that was resulting from the base deflection. Fortunately, this was a through hardened gear set and the teeth "wore in". If it had been case hardened, it likely would have cracked a tooth.

the manufacturer had made an error in assembly, but we then found that the torque on the base was flexing the I-beams by over 4 mm (0.18″)! In two other examples:

- In a processing plant with a large conveyor with a 200 mm (8″) wide open gearset, we had to intentionally misalign the gears by 2 mm (0.080″) so there would be relatively full contact across the teeth when the conveyor was running loaded!
- In an injection molding plant with a series of large reducers, we were in their shop when they showed a static contact check with bluing that looked great. Then we asked them to clean off some teeth so we could all see the actual wear pattern and it was pretty evident that it was about half-way across the teeth. (Shimming one side of the box by 2 mm (0.080″) corrected the misalignment.)

A gear manufacturer we work with has an easy solution to checking actual tooth contact. In their shop, they check the static alignment by putting a very thin coat of blue layout dye on the teeth and then run the reducer and check and record the contact pattern. Then, after that passes inspection, they ship the unit with red layout dye on the teeth.

Bases and housings of all sorts of machinery deflect because of the forces on them. In these applications checking alignment or tooth contact patterns during a shutdown can be very misleading because the deflection force isn't present. Maintenance guides should spend a lot of time talking about the benefit of precision alignment, but we also have to beware of *dynamic soft foot*.

PHOTO 11.6 This is the drive pinion from a haul truck wheel motor and it really emphasizes what happens with structural deflections. These haul trucks were pulling heavy loads up out of a mine, then the 300 +/– ton trucks were coasting downhill to be refilled. The difference in wheel motor stresses flexed the housings to the point where the two sides of the teeth couldn't both be in perfect contact. We often see similar alignment challenges in mobile equipment such as draglines and bucket loaders.

With through hardened gears misalignment such as that in Photo 11.5 frequently appears as pitting after only a short time in run. With a surface hardened gear, such as the haul truck drive pinion shown in Photo 11.6, there should be no visible damage and it is more difficult to see. In this example, the contact pattern that ends well before the edge of the pinion tooth shows that the peak contact and bending stresses are well above two times the ideal design stress.

The emphasis on contact patterns is needed because they are tell-tales of the actual forces on the teeth during operation. Looking at the bending stress that causes tooth breakage we see that it is a direct function of the load, i.e., if the transmitted power increases by 30%, the bending moment increases by 30%. However, the Hertzian fatigue loading is much more complex. For roller bearings where the fatigue stress's effect on life is an exponential factor, we know that doubling the load on a roller bearing cuts the life by a factor of 10. From what we have seen, this also applies to gear tooth deterioration, but it may be conservative, i.e., doubling the load reduces the life by more than a factor of 10.

Other than coating the teeth with a layout dye, the best way we have found to monitor the actual contact is to look at the gears in operation using a strobe light and an infrared scanner, watching the tooth appearance and the temperature distribution both across the teeth and around the entire gear. On the large open gear sets we try to keep the variation to less than 6°C (~10°F). This inspection is easy to do on an open gear set but almost impossible in an enclosed unit. However we have run enclosed units, removed the bolts on the inspection plate until only two or so are left, then shut the unit down and quickly opened the cover, taking the temperature readings as soon as the unit stops moving.

When we started the Reliability program at the plant where I worked, we had no idea of the value of good alignment. As our alignment techniques improved the life of our gears increased amazingly and our gear cost per ton of product decreased measurably.

BACKLASH

One of the definitions earlier in this chapter states that backlash is "The effective clearance between the mating teeth". Measuring the backlash between a set of gears is fairly simple. One of the gears is locked in position so it can't move, then the other gear is rotated back and forth and the clearance between the two meshing teeth is measured with a dial indicator or feeler gage. With case hardened gears the backlash typically changes little over the life of the gears because there is usually little measurable wear but with through hardened gears the backlash continually increases.

If the gears always drive in one direction this increase is generally not very important because the load is continuous, and the increase is really a measure of the tooth wear. With reversing gear sets, wear and increasing backlash can become a source of impact loading and must be evaluated. With the mine hoist gears shown in Photo 11.8 the increase in backlash had reached more than 5 mm (0.2″) but that wasn't a problem because the motor controller started the system slowly and there was very little impact force involved.

Design Life and Deterioration Mechanisms

Earlier we mentioned how:

- AGMA standards for reducer design call for an L_{10} life of 5,000 hours, only 5/8 of a year, and we expect any reducer to last far longer than that.
- The way they compensate for what the customer actually wants is to change the service factor, i.e., the input power rating of the reducer.
- In the 1960s it wasn't unusual to have a 2.8 SF rating but, because of competitive pressures SFs less than 1.5 are frequently seen.

Compensating somewhat for that is the fact that designs are much more precise.

The AGMA categorizes gear failures into categories of wear, surface fatigue, plastic flow, breakage, and associated gear failures. (We approach the failure analysis procedure from a different perspective but think their literature on gear failures is an excellent source of information.) Looking at their classifications of gear failures, they include:

1. *Wear* – The term "wear" covers a huge range of deterioration rates from mild polishing of gears that may last more than a hundred years to adhesive and abrasive wear that can render the gears essentially useless in a few

minutes. For through hardened gears most wear is predictable and can be monitored. For case hardened gears any wear more substantial than polishing is usually a sign of impending failure.

2. *Surface fatigue* – This includes pitting and spalling and again, from a practical standpoint, surface deterioration of case hardened gears is worrisome.

3. *Plastic flow* – With surface hardened gears this includes rippling, a relatively slight deformation of the gear surface. Through hardened gears frequently suffer from rolling and peening that leave the surface profile badly damaged.

4. *Fracture* – In addition to the classical fatigue and overload fracture mechanisms we've covered in earlier chapters, case hardened gears sometimes suffer from case crushing when the contact loads are so great that the underlying structure deforms, cracking the hardened case.

THROUGH HARDENED GEAR DETERIORATION MECHANISMS

Pitting and Wear

Gear deterioration can be very confusing and complex. Through hardened gears are frequently not very hard and not very precise and they tend to pit and wear in a relatively predictable manner. However, with case hardened gears, any surface distress is cause for concern and pitting is often a sign of an impending disaster. We suspect that this huge difference in operating appearances leads to many misunderstandings about through hardened tooth wear and probably results in more unneeded gear replacements than anything else

A good side of through hardened tooth deterioration is that it results in relatively predictable degradation and it isn't uncommon to see through hardened teeth wear substantially before there should be any concern about failure. We have seen several smoothly operating older reducers where less 20% of the original tooth thickness remained yet they were still doing their jobs and in many through hardened applications the condemning thickness is typically around 60% of the original tooth thickness and sometimes even less. The gear shown in Photo 11.7 had run for 11 years in a large printing press. It was removed from service but was still functioning well with more than 50% of the tooth gone. (A careful inspection of the photo also finds that this relatively soft through hardened gear was very poorly aligned)

There are a number of through hardened tooth deterioration mechanisms including pitting, adhesive wear, rolling, and peening but most gear wear starts with pitting and, with the combination of loads, materials, and lubricant film, it is frequently unavoidable. Figure 11.10 shows three general classifications for pitting and their common locations.

Referring back to Figure 11.4, with an understanding of the relative tooth motion, exploring each of these areas finds:

1. *Corrective pitting* – As two gears mesh together the high spots rub against each other and Hertzian fatigue causes small spalls. As time goes on, more and more of the tooth surfaces contact each other and the wear pattern becomes wider and wider. The pits should be relatively small, and generally

PHOTO 11.7 A badly worn "through hardened" gear that actually wasn't hardened at all. Testing showed it was machined from SAE 1045 steel and was about BHW 180.

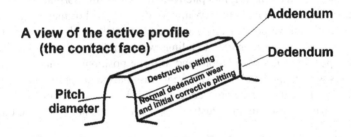

FIGURE 11.10 The areas where pitting of through hardened gears is usually first seen.

not exceed 1.5 mm (~0.06″) in diameter – but see the note below. They support the development of a lubricant film and are in the dedendum of the teeth. Corrective pitting starts rapidly, but as the contact improves the wear rate decreases, eventually reaching a low but continuous level. (There really is a difference between wearing in and wearing out!)

2. *Normal dedendum wear* – This occurs in the dedendum of the teeth as a long-term function of the tensile fatigue stresses. It generally shows up as the gears age and the deterioration rate is slow and readily monitored.

3. *Destructive pitting* – These pits are large and irregular and don't support the lubricant film. They begin in the dedendum but rapidly progress into the addendum with the resultant tooth wear and noise.

Gear texts usually speak of corrective pitting being 1 mm in diameter or so but the dividing line between acceptable corrective pitting and worrisome destructive pitting is not very clear and, from a practical standpoint the pitting to be concerned about varies with the speed of the gears and the lubricant. The pits shown in Photo 11.10 are much larger than those shown in Photo 11.8 (and more than 3 mm [1/8″] wide in some cases) yet both gears will last more than 20 years. The gears in Photo 11.8 are in an enclosed reducer and fed with a pressurized and filtered spray using an ISO 320 gear oil. Those in Photo 11.10 use a heavy open gear grease and the pits will grow smaller once the pattern extends all the way across the tooth.

One of the difficulties in assessing through hardened tooth wear occurs on start-up when the teeth begin pitting and it is initially difficult to tell the difference between corrective and destructive pitting. Figure 11.11 shows a chart of the pitting rates in a typical gear installation but sometimes monitoring of pitting can be difficult. In the past, carbon paper was run between the teeth to measure tooth mesh changes but we generally use a combination of oil sampling, digital photography, and measurements with a gear tooth micrometer. The figure shows the difference showing up over a few weeks and, in a typical industrial application with an enclosed reducer running 24 hours per day, if the wear rate doesn't start to decrease within four weeks you probably have a destructive wear situation that will only get worse.

An example of both how pitting changes with load and the predictable nature of pitting can be seen in Photo 11.8. This is a mine hoist gear that lifts a personnel elevator over 0.7 km (~2300 ft) and it was originally driven by a 750 kW (1000 hp) motor. After about 15 years of operation they decided to increase the motor to 1125 kW (1500 hp). Based on the original pitch line thickness, the normal dedendum wear rate from pitting had been about 0.05 mm/yr (0.002″/yr). For many years after the motor was replaced we periodically measured the tooth width just below the pitch line using a *gear tooth micrometer*. Readings were carefully taken once or twice per year in six positions across the tooth and at three positions around the gear. Then the readings were logged and the tooth wear was charted at approximately 0.2 mm/yr (0.008 in/yr). With a calculated condemning thickness well in excess of what would ordinarily be used in an industrial application, this not only gave the mine a

PHOTO 11.8 A mine personnel hoist gear where we routinely tracked the wear using a gear tooth micrometer.

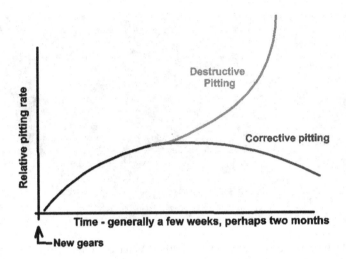

FIGURE 11.11 Corrective pitting should gradually grow across the width of the through hardened teeth and then the pitting rate should decrease.

predictive maintenance tool that they could use to justify replacement gears, it also prevented an unnecessary early expenditure. (In this example, increasing the available motor power by 50% increased the wear rate by about a factor of 4.)

The gear tooth micrometer is essentially two sets of vernier calipers that are joined together. The original one we used is shown in Photo 11.9 and we now use a digital one for improved accuracy. For field measurements, the depth was set about 10% below the pitch line, we then measured and charted the tooth width. (One caution about using the micrometer is that you have to be sure there isn't a lip from plastic deformation at the top land of the tooth.)

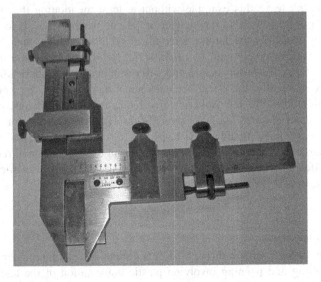

PHOTO 11.9 Our original gear tooth micrometer. Now replaced with a digital version.

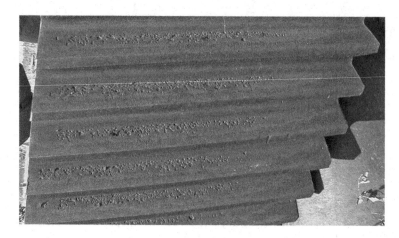

PHOTO 11.10   An example of corrective pitting. As the pits grow across the teeth and the contact extends over more and more area, the wear rate slowly decreases. Unfortunately, in this example, the rust will decrease the gear life.

Photo 11.10 shows a through hardened gear that has been removed from a large piece of mining equipment. The teeth show some dedendum wear and some of the pits are larger than we would like to see, but it had been in run for eight years and a decision was made to change it despite the fact that it has many years left on it. Looking at this gear we can see that the pattern has not extended all the way across the teeth and the tooth profile is still in very good condition. This is an ideal application for periodic inspection with the gear tooth micrometer and this gear should be carefully stored and preserved as a viable spare.

We shouldn't leave this example without at least mentioning the subject of corrosion of gear teeth. In the chapter on fatigue we discussed at length how corrosion decreases the fatigue strength of metals. With through hardened teeth that are relatively soft, less than about HBW 180, corrosion will basically accelerate the pitting action and will result in shortened gear life. As the teeth become harder the effect becomes much more serious. We know from numerous studies that atmospheric corrosion can cause the fracture of highly stressed materials with hardness levels below HRC 40. Our rule of thumb is that gears in storage should always be protected from corrosion. In addition, on a case hardened gear, any visible rust is proof of damage that will reduce the fatigue life and visible heavy rust has probably already resulted in microcracking of the case.

Plastic Deformation

Investigating other common through hardened tooth deterioration mechanisms, we see both rolling and peening involving plastic deformation of the teeth. In these examples, the stresses are high enough to physically deform the teeth. *Rolling*

PHOTO 11.11   The gear on the left shows rolling where the tooth metal has been pushed up over the top and over the sides of the teeth. It's also heavily worn.

involves contact stresses large enough to physically move the metal up over the top or around the side of the tooth resulting in a sharp ridge over the top land of the tooth or a mushroomed appearance on the side. *Peening* is the hammering of the mating teeth resulting in a deformed (peened) tooth profile. In both, either the lubrication has been inadequate or the loads have been excessive and with continued operation the gears will grow progressively noisier and rougher.

Looking at the two bevel gears in Photo 11.11, the one on the right is like new while the one to the left has lost a significant portion of the tooth width. Wear has thinned the teeth and rolling has moved the metal around the edges of the teeth but in a heavily loaded low speed application the unit still has many more productive hours left in it. (Looking at the rolling damage, this is a reversing gear and as long as the impact loading from the increased backlash isn't excessive, we expect this gear would have traveled about 2/3 of the way through its life.)

Adhesive Wear

There are other terms commonly used to describe adhesive wear with scuffing, scoring, galling, and seizing being the most frequently heard. Adhesive wear involves breakdown of the lubricant film and momentary welding of the teeth, followed by tearing a particle out of one of the teeth. It is most often seen at the tips and the roots because those are the areas where it is most difficult to develop an effective lubricant film. Photos 11.12 and 11.13 both show damage from adhesive wear.

Photo 11.12 is a close-up of a worn tooth and the radial marks from poorly lubricated contact with the opposing teeth are plainly visible. The wear rate decreases as the contact surface approaches the pitch line. At the pitch line there is pure rolling and both there and above it there is a greatly improved lubricant film and decreased wear rate. (If the contact stress had been lighter this "scuffing/scoring" would be

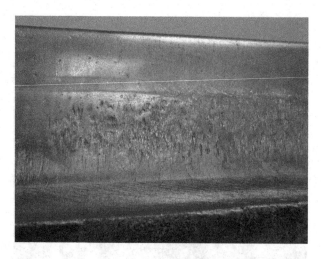

PHOTO 11.12   A close up of adhesive wear of a through hardened tooth.

PHOTO 11.13   A much more serious example of adhesive wear. In this example, a bearing on this reversing application reducer failed and put an extreme load on the teeth and lubricant. The lubricant couldn't do its job and result is severe adhesive wear and rolling of the teeth material.

likely be replaced by polishing and the metal removal rate would have dropped substantially.) This is a graphic example of the effect of the combined rolling and sliding action.

Tooth Fracture

Gears fail from fatigue much as the other mechanical components and Photo 11.14 is an interesting example.

PHOTO 11.14 A cast steel through hardened gear that was relatively new. Inspection of the fracture face can tell us that the tooth was poorly aligned and that it failed from fatigue (repeated loads). Also, the extremely large instantaneous zone tells that the final fracture load was extreme.

CASE HARDENED GEAR DETERIORATION MECHANISMS

Earlier we said that case hardened gears should show essentially no wear or deterioration and, our experience has shown that for most of these gears this is true. However, there are occasional trouble areas and warning signs that should be recognized.

One caution we should mention is that, because case hardened gears wear more slowly than through hardened gears, there has been a tendency for some engineers, when an older gear or a reducer wears out, to specify case hardened replacement gears. If the foundation for the replacement is sound and stable and the replacement designed for the peak loads, the assembly will run for years. But if the support structure isn't sound and the case hardened teeth end up with a distorted contact pattern (dynamic soft foot) the replacement will fail rapidly.

Micropitting

Micropitting occurs when a surface is heavily loaded and has the appearance of a lightly sandblasted surface. (Some people call it "frosting" but micropitting is a more accurate descriptive term.) In some instances micropitting is of absolutely no concern while in others it is a danger warning. Photo 11.15 shows a heavily loaded 750 kW (1000 hp) reversing drive from a steel mill with obvious diagonal patterns across both sets of teeth. These patterns are a series of very fine pits and in this case the micropitting:

- Occurred because of the slight irregularities on the machined surface
- Is absolutely nothing to be concerned about
- Shows that the gears are heavily loaded
- Shows that the alignment is excellent

PHOTO 11.15   If you could look at the micropitted lines on these teeth under high magnification, you would see a series of extremely fine pits, as though they were sandblasted.

In many situations however micropitting is cause for concern as is shown in the two examples below. Photo 11.16 shows a 3.2 cm (≈1.25 in) wide centrifugal air compressor drive gear with micropitting on the left side of the teeth while Photo 11.16 is of a gear from a small generator drive that shows micropitting on the right side. In both applications the fact that the attack is concentrated on one side and is relatively obvious indicates serious alignment problems and, looking at Photo 11.17, the

PHOTO 11.16   This centrifugal compressor drive gear was installed improperly and was misaligned with a heavy load on the left side of the teeth.

PHOTO 11.17 This is micropitting on the upper two teeth followed by a pit on the lower one. This sort of micropitting, i.e., when it is on one side of the gear, should ALWAYS be a cause for concern.

micropitting has developed into macropitting and there is a pit on the lowest tooth in the photo.

One of the design requirements is that surface hardened gears must be well aligned. The elasticity of the teeth will correct for minor misalignment and micropitting along one side of a gear is a warning that there are both very high local loads and excessive misalignment.

Rippling

The appearance of *rippling* on a case hardened gear falls in the same basic analytical category as micropitting. It tells that the gear has been very heavily loaded and there is some plastic deformation of the surface. Again, as with micropitting:

1. If the rippling is consistent across the tooth it isn't as worrisome as if it is on one side of the tooth. It is a sign that the gear should be closely monitored but not an indication of an immediate concern.
2. If the rippling is on one side of the tooth it is a tell-tale that there is misalignment, a warning that a failure may occur in the future, and a call for regular inspections to ensure that there isn't further deterioration.

Pitting

With the knowledge that micropitting shows there are heavy loads, any *macro* pitting is source of serious concern. Having said that we should emphasize that pitting can't automatically be considered a warning of an impending disaster, but it does have to be carefully analyzed along the lines of any other failure (Photo 11.18).

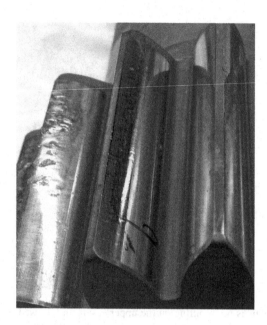

PHOTO 11.18 The tooth on the far left shows rippling, a situation in which the load is heavy enough to deform the hardened case and an indication that the gear is very heavily loaded. Also, the loading on one side is proof that there is an operating misalignment.

Our logic about being concerned about pitting of surface hardened gears goes back to the tensile strength of the teeth. When the case is HRC 55, its tensile strength is approximately 1.9 GPa (275,000 psi). If the core is HRC 35, its tensile strength is less than 2/3 that of the case. If the load is heavy enough to break the case, how long would a core that is only 60% as strong?

One of the problems with this approach is that to some degree we are comparing apples with oranges. The case damage occurs because of Hertzian fatigue, with complications from sliding, while the tooth breakage results from cantilever loading. Nevertheless, probably 99% of the surface hardened teeth we've seen that have failed from low or high cycle fatigue have had pitting problems. In the analysis of pitted surface hardened teeth, look carefully at the shape of the pits. Oblong pits, with their long axis along the length of the tooth, and relatively deep sharp bottoms are serious stress risers. On the other hand shallow, round pits with round bottoms are nowhere near as problematical.

In Photo 11.19, we see most of the active profile a case hardened tooth. The bottom of the photo ends at the root radius and the pits are starting to become elongated and link together, a source of concern. In addition, careful inspection shows some deterioration on the right side of the tooth and faint indications of rippling on the left side. Any recommendation on the future of the gear would depend on how long it

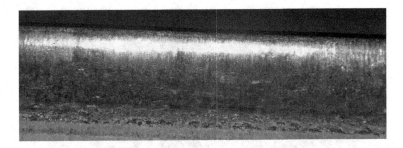

PHOTO 11.19 Showing some pitting in the dedendum of the tooth with rippling above it.

PHOTO 11.20 A view of a spall on a case hardened pinion caused by improper tempering that left a high residual stress between the case and core. (This is similar to what caused the transmission gear failure mentioned early in the chapter.)

had taken to reach this state however our opinion is that we would be very reluctant to run it in any sort of a critical drive.

Spalling

Small pits are an indication of high localized stresses and appear on heavily loaded gears. Large spalls are indication of metallurgical problems in the manufacture of the gear. Photo 11.20 is a bit of a difficult photo to view because it was taken through an expanded metal guard however, in the center of the photo is a spall that covers more than half of the active flank of a 3.8 cm (~1.5″) tooth. The fact that there is this huge area of separation between the case and the core is almost always evidence of a heat treating problem.

Case Crushing

Photos 11.21 and 11.22 are both of some teeth that suffered from case crushing, sometimes called sub-case cracking. It happens when the load on the tooth is so great that it causes plastic deformation of the softer core. But the case is harder and less ductile than the core, can't tolerate the deformation, and fractures. The first photo

PHOTO 11.21 This is a close-up of a large tooth and the two cracks in the center of the picture are from case crushing, i.e., the brittle case cracked because the softer core deformed.

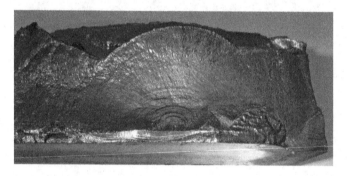

PHOTO 11.22 This is another example of case crushing, but this one clearly shows the fatigue crack starting inside the tooth. Looking to the right side, the full case thickness is visible and this gear was badly worn.

shows a typical view while the second shows how the fatigue crack started within the tooth.

Case crushing is not uncommon in mining and other industries where gears are frequently run to destruction.

Thermal Power Rating

As you know, there are inefficiencies in every gearset and the normal result is that they generate heat and, as the operating temperature increases, the viscosity of the lubricating oil decreases. The thermal power rating of a gearbox is the manufacturer's guideline as to the maximum allowable power transmitted without adding supplemental cooling with continuous (more than three hours) operation.

Important is that this rating is based on a clean housing and typical plant dirt will reduce heat transfer and increase the lubricant temperature. Critical are that:

- The oil viscosity is adequate at the operating temperature.
- The elevated temperatures don't result in additive depletion between oil changes.

ANALYZING GEAR FAILURES

Approaching a gear failure, the goal should be to understand why it happened and the analysis should be done in the same basic manner as any other failure analysis. The difficulty with a gear failure analysis is that they are very complex and it is not easy to separate some of the mechanisms. However, start by categorizing the type of failure as shown in Figure 11.12, then look at the various features, i.e., the last two tiers of this chart.

Some ideas that should be used as a starting point for a field analysis are:

1. *Look at the surroundings* – What is the support structure like? How rigid is it? If it is an open gear set, how is the gear mesh adjusted? How true does the driven gear rotate?
2. *Load* – How do the operating loads compare with the design loads?
3. *Load distribution (or alignment)* – Are the gear and pinion contact patterns uniform all the way around both and across the teeth?
4. *Temperature and lubricant film thickness* – How do the operating temperature and the lubricant package compare with the original specifications? If the loading has changed, has the lubrication also changed appropriately?
5. *Gear tooth geometry and finish* – How does the current tooth profile (actual tooth shape) compare with the original gear specifications? Was it in storage and allowed to rust?
6. *Maintenance* – What is the maintenance history and what was the last maintenance performed on the unit? Who conducts the maintenance? Is it done by plant personnel or contract personnel, and how experienced are they?
7. *Gear metallurgy* – Is it through hardened or case hardened? When there are large spalls, always ask for a metallurgical examination.

If the gear has been removed from the application and you can look at it in a laboratory or clean shop site, there are lots of other questions to be asked and they are outlined below.

Bending fatigue cracking indicates the cantilever strength of the tooth has been exceeded over some time. The questions are designed to help find why the load was excessive.

1. For all gears
   A. Is it a reversing application?
      (1) If so, what sort of impact occurs when the load is reversed?
      (2) Was there wear on the teeth? What was the backlash?
      (3) Does the drive have a soft start on reversing?
   B. Where does the fatigue crack start? Is it at the center of the teeth, showing good alignment, or toward one edge?
   C. Are there fatigue cracks on many teeth, indicating a very high overload, or only on one or two teeth, showing the loading was only slightly above the fatigue strength? (In answering this question consideration

Gear Failure Analysis

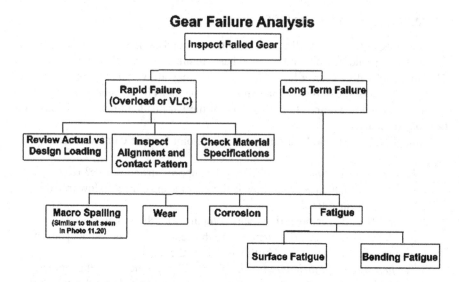

FIGURE 11.12   A simple logic tree for starting on a gear failure analysis. Depending on the category of the failure, refer to the earlier chapters and the suggested questions below for a solution to the physical causes.

has to be given to the time since the first fatigue crack originated. There may be many fatigue cracks because the gear has been forced to run for a long time since the cracking started.)

D.   Where is the contact pattern? Has it changed recently? Is there any evidence of contact on the unloaded side of the teeth? Is there obvious wear of the teeth?

(1)   In looking at the pattern, was the root clearance excessive?

(2)   For case hardened teeth, variations in the wear pattern may be difficult to see and it may be necessary to slowly rotate the gear with a bright light on the teeth.

(3)   For through hardened teeth:

- Some wear is expected but it may have gone too far.
- Is the wear uniform across the teeth and around the gear? (If it isn't uniform, that shows the gears are either misaligned or mismachined.)

E.   What was the original material specification? What is the hardness specification and what is the actual tooth hardness?

F.   What is the condition of the foundation?

2. For case hardened teeth

A.   Is the any evidence of micropitting? Is the pattern skewed, indicating eccentric loads or misalignment?

B.   Is there any evidence of plastic deformation, i.e., a thin ridge along the top of the tooth?

C.   Did the cracks originate at pits, evidence of very high loads?

(1) Was the lubrication as specified? What load was the gear origi-
    nally designed to carry? What was the service factor? How does the
    load vary with time? Do you have sound verification of the loading
    variations?

(2) Was the gear ever allowed to rust and corrode, greatly reducing the
    fatigue strength?

Surface fatigue – Hertzian fatigue spalling, pitting, and micropitting happen when
the surface loads are excessive; a frequent contributor to this situation occurs when
the lubricant film is inadequate.

A. Where does the spalling/micropitting start? Is it at the center of the teeth
   (showing good alignment) or far toward one edge?

B. Look at the mating gear. What is the tooth profile like? Is it in good
   condition?

C. The correct lubricant film should prevent pitting. What has the lubrication
   history been and how are the lubricants applied? Are EP lubricants used?
   Do the change/application intervals recognize the possible deterioration of
   the EP lubes? How does the lubricant operating temperature compare with
   the design operation temperature? What have lube samples shown?

D. Is the surface fatigue pattern uniform around the gear, indicating good gear
   geometry, or is it inconsistent, indicating machining errors?

   (1) For through hardened teeth, is the wear uniform across the teeth and
       around the gear? (If it isn't it shows the gears are either misaligned or
       mismachined.)

   (2) With case hardened teeth:
       – Some light and uniform micropitting is not of concern. But, with
         the alignment needs of hardened materials, if the pattern is skewed
         or erratic it indicates a potential problem.
       – Eccentric micropitting, on one side of a tooth, is an indication of
         abnormally high surface stresses.
       – The presence of pitting on through hardened gears is another result
         of abnormally high surface stress.
       – An isolated large spall (or spalls), greater than 6 mm diameter (1/4″)
         is almost always an indication of manufacturing problems.
       – Was the gear ever allowed to rust and corrode, greatly reducing the
         fatigue strength?

FAILURE EXAMPLES

INTERMEDIATE PINION FAILURE EXAMPLE 11-1

This was a case hardened intermediate pinion on a reducer that had run for about a
year. This is a three-stage reducer that was removed from service because of a vibration analysis indicating a problem. It had never been apart and the loads and speeds
were well within the design specifications.

PHOTO 11.23   Spalling on an intermediate pinion.

Inspection found:

- Another spall about 120° around the case hardened intermediate pinion from where Photo 11.23 was taken.
- The lubricant tests indicated the lubricant was as specified by the manufacturer.
- Other than the two spalls there was no evidence of other damage or misalignment.

Conclusions and recommendations – There is nothing the user can do to cause this type of spalling in an intermediate gear set. We suspected a metallurgical problem and recommended a detailed metallurgical analysis to determine the true physical roots. (This found there was a processing problem.)

BROKEN GEAR TOOTH – FAILURE EXAMPLE 11-2

This was a single tooth from a large slow gear in a materials processing plant. It had run for about six months at 5 rpm when the tooth failed.

Inspection of the fracture face found, as shown in Photo 11.24:

a. The tooth has a relatively thick case tempered to about HRC 52. (We have blackened the case surface in the bottom center of the photo.)
b. The tooth failed from fatigue but, knowing that it has a relatively fine grain structure, the failure was very rapid.
c. The load was poorly centered and the crack started far to the right side of the tooth and the left half of the tooth was the instantaneous zone.
d. The wear pattern indicated the root clearance was excessive.

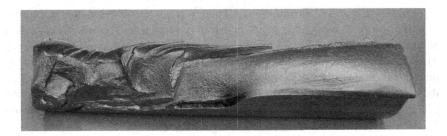

PHOTO 11.24   A case hardened tooth.

Inspection of the contact surface of the tooth found some pitting which was also centered far to the right side of the tooth.

Conclusions and recommendations – The gears were seriously misaligned, and our recommendation was to correct the misalignment and the center distance. (This is another example of the sensitivity of case hardened gears to misalignment.)

Large Pump Drive Gear – Failure Example 11-3

Photo 11.25 is of a 2625 kW (3500 hp) gear out of a pump drive that has two 1310 kW (1750 hp) 1800 rpm motors driving pinions that meshed with this gear. The output speed was about 400 rpm. This is a flame hardened tooth with a proprietary contour and the heat-treated profile is readily visible on the side of the teeth. The drive has been in operation for more than a year and there have been several comments from technical and maintenance personnel about gear noise.

Inspection of the fracture face shows that the gear is driven to the left while the river marks tell us that the crack growth is from left to right, against the drive forces. An inspection of the tooth surfaces shows there has been repeated contact on the "unloaded" side of the teeth.

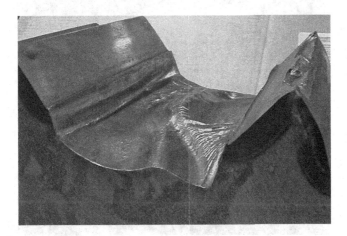

PHOTO 11.25   The pinion from a 2625 kW (3500 hp) pump drive reducer.

Conclusion and recommendations – There was a torsional resonance in the gear teeth, i.e., the teeth were being struck or driven at a frequency that would excite their natural frequencies (confirmed by vibration analysis). The reversing forces put the weakest section of the case into tension, causing a crack that then propagated in the much weaker core section. As shown earlier in the chapter, for there to be contact on the unloaded side of the tooth, the tooth driving forces had to be more than twice the design forces. Changing the natural frequency of the gear and shaft assembly would be extremely expensive and we recommended they retire the reducer.

HAUL TRUCK PINION – FAILURE EXAMPLE 11-4

Photo 11.26 is of a pinion from the rear wheels of a 300+ ton haul truck. This drive axle drives the truck uphill and is loaded on the other side by the braking system in the downhill run and it had been in operation about six months.

Inspection found this tooth shows two large spalls and another tooth out of the photo showed a single large spall near the left end of the tooth. There is no visible contact pattern in the photo and the actual contact pattern looked good on the more heavily loaded side.

Conclusions and recommendations – The large size and unusual locations of the spalls lead us to suspect a materials problem possibly complicated by misalignment. We recommend:

- Carefully reinspecting this gear using wet fluorescent magnetic particle testing to determine if there is any more cracking. (This may show a more definite pattern.)
- Having a metallurgist section the tooth, analyze the metallurgy and the heat treatment.

PHOTO 11.26   The pinion from a 300-ton haul truck.

PHOTOS 11.27 AND 11.28   Two pictures of gears inside a large reducer.

PAPER MACHINE REDUCER GEAR – FAILURE EXAMPLE 11-5

This reducer had been on the paper machine for over ten years but the gears had been replaced about four years before the failure. Shortly before the reducer failed the cover had come off the driven roll, requiring a shutdown. When the machine was restarted they noticed the reducer was very noisy and they replaced it.

Inspection found three broken teeth on this through-hardened gear. The oldest, based on the appearance of the fatigue crack surface, is shown in Photo 11.27 and the other broken teeth were spaced around the gear. Further inspection of the gear showed that it had been misaligned and wearing irregularly as shown in Photo 11.28. There were some larger pits but most of the pitting was relatively fine and a combination of corrective pitting and normal dedendum wear. The historical data on the reducer lubricant appeared reasonable and hardness testing of the gear showed it to be HRC 35. Gear tooth micrometer measurements showed they had worn by about 8%. (There was no drive motor load data to look at.)

Conclusions and recommendations – The wear pattern shown in the second photo shows these through-hardened gears were obviously misaligned, but they should not have failed with only 8% of the tooth missing. The surface of the fatigue crack indicated that the gears had only run only a short time between initiation of the cracking and the final fracture. We could also see that the pieces of the first tooth failure then caused other damage; however, the first fracture shows there was a very heavy load on the tooth. Based on the above, our conclusion was that the failure of the cover increased the reducer load for a short time and resulted in the gear failure. We also recommended they contact the reducer manufacturer to discuss the internal misalignment and possible correction. (We suspected a defective foundation.)

KILN REDUCER INTERMEDIATE GEAR – FAILURE EXAMPLE 11-6

This kiln had been in operation for over 20 years when the management considered increasing the motor horsepower by 10% to allow the kiln to rotate faster. The reducer gears had not been a problem and they were not certain they would increase the kiln speed. The primary interest was that they wanted to be certain that there would be no problems over the next ten years.

Inspection – This was a three-stage reducer and the first five gears looked like the day they were manufactured. The tooth profiles showed no wear and the contact patterns showed all the way across the gears and were consistent around all five gears.

PHOTO 11.29 Looking inside a large reducer with an oil pipe obscuring some of the teeth.

The low speed output gear had 63 teeth and typically rotated at 20 to 24 rpm while the kiln was running. Inspection of this gear showed irregular pitting on several of the teeth as shown in Photo 11.29. (That is an oil spray pipe across the center of the photo.) On some of the teeth there was no pitting while on others it extended more than half way across the tooth. There was no evidence of plastic deformation. Hardness testing of the teeth showed them to be HRC 31.

Conclusions and recommendations – The low speed gear shows corrective pitting, a result of minor machining errors while the gears were being manufactured. With this little wear over the last twenty years we would expect the reducer to easily last more than another 20 years even if the horsepower were to be increased by 10%.

INBOARD–OUTBOARD PROP DRIVE GEAR – FAILURE EXAMPLE 11-7

This wasn't the subject of a lengthy investigation. It is the gear off an outboard drive on a boat that is frequently used by a friend to go out to sea for sport fishing. Eventually it failed and they asked us to look at the gear. There were other problems, but two interesting points can be seen looking at Photo 11.30.

  a. On the teeth to the left, there is plastic deformation of the contact surface.
  b. Then looking in the center of the photo, there is a small but obvious ridge of plastic deformation along the top lands of the teeth.

We asked if they regularly checked and changed the lower unit lubricant. The answer was a sheepish "no", but they promised to be more conscientious in the future.

REVERSING INDUSTRIAL DRIVE GEAR – FAILURE EXAMPLE 11-8

This is not the typical failure analysis, but is a valuable lesson. After hearing the plant's description of what had happened and inspecting the contact pattern, it

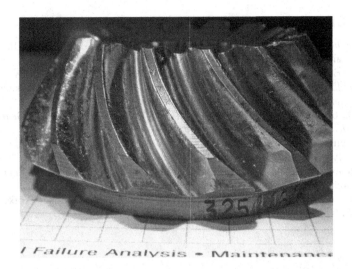

PHOTO 11.30   Lower unit gear off an off-shore fishing boat.

PHOTO 11.31   A relatively easy failure analysis, i.e., it broke instantaneously from an overload! Interesting is that the chevron marks show the fracture began about 20% of the way from the left side and prove that the alignment was very poor.

was evident that severe misalignment played the major role in causing the failure. However, the fracture photo below is a great example of the value of a simple field inspection. We've used a flashlight off to one side and shadowed the fracture face, similar to what was shown in Figure 4.3. From that it can be seen that:

- This was a brittle (instantaneous) failure.
- The gear is case hardened. (Notice how the outer layer is much smoother than the core.)
- The chevron marks are obvious and they point to the failure starting almost exactly one-fourth of the way from the left side of the tooth, proving that the alignment was very poor (Photo 11.31).

Summary – Follow the outline and look at every symptom. The gears will tell you why they failed.

BIBLIOGRAPHY

ASM Handbook (8th edition), Volume 1, Edited by Taylor Lyman, 1967; Volume 10, Edited by Howard E. Burger et al., 1975, ASM International, Metals Park, OH.

Avallone, Eugene, Baumeister, III, Theodore, and Sadegh, Ali M., *MARKS Standard Handbook for Mechanical Engineers*, McGraw-Hill Education, New York, 2006.

Gear Technology, Numerous articles and issues, Randall Publications LLC.

Parrish, Justus, *Factors Affecting the Properties of Carburized and Hardened Gears*, printed in *Metal Progress (ASM International)*, 1983.

Sachs, Neville W., *Failure Analysis Made Simple: Bearings and Gears*, Reliabilityweb.com, Ft. Myers, FL, 2015, ISBN: 978-1-941872-30-7.

Source Book on Gear Design, Technology, and Performance, Edited by Maurice A.H. Howes, American Society for Metals, Metals Park, OH, 1980.

Walsh, Ronald, *Electromechanical Design Handbook*, McGraw Hill, New York, ISBN-10: 0071348123.

12 Fastener and Bolted Joint Failures

When we conducted training programs at the big, wet, dirty and caustic chemical plant where I worked, the first workforce classes involved fasteners. There were some truly impressive disasters with a variety of bolt failures, so the emphasis in those early classes was on preventing bolt breakage. Eventually, we sat back and did some more detailed analyses of our situation and realized that the major physical cause of most of our leakage problems was poor bolting practices. Since that time, we've worked with hundreds of plants across North America and invariably we find that a major (if not *the major*) maintenance cost is repairing leaks. Flange leaks don't just happen, and those steam and process leaks that required expensive cold shutdowns to repair are almost always unintentionally caused by the people that put them together.

We (and many others) have also spent a fair amount of time looking at the distortion and other problems that poor bolting practices can cause and how that relates to reliability problems. My opinion is that careless bolting procedures probably cause more unrecognized damage that anything other than corrosion.

In this chapter, we hope to:

1. Provide data on the common fastener materials available.
2. Help you to understand how bolts work and how they fail.
3. Explain the importance of uniform tightening procedures, emphasizing that proper flange assembly requires intelligence, care, and skill.
4. Help you to diagnose fastener and bolted joint failures.

HOW BOLTS WORK

The idea of bolt elasticity is a critical one and a bolt acts a little like a very strong rubber band as it is being tightened. This stretching and the resultant clamping force then hold the parts together. In Figure 12.1, the sketch of a typical bolt includes the *grip length* which is important because it is the stretch in this area that provides the elasticity that does the clamping, the property critical to maintaining effective bolting performance.

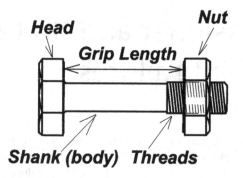

FIGURE 12.1 A typical nut and bolt assembly with some important terms.

With equal loads, all grades of steel stretch the same amount and the benefit of going to a stronger bolt is that the elastic range of the steel is greater and the bolt can stretch more before it plastically deforms. Putting that into practice we see:

(Remember, though, "There is no such thing as a free lunch". The tradeoff for increased strength is increased susceptibility to damage from corrosion and reduced toughness.)

With a conventional nut and bolt assembly, as it is tightened and the assembly develops this clamping force, some of the tightening work is absorbed by the stretching while the rest of it has to overcome friction, either between the rotating piece and the mating threads or between the rotating piece and its contact surface. Therefore, with better lubrication and less friction, more of the tightening torque can go into developing the clamping force. There have been many studies of this area and the current thought is that, on average, about 40% of the turning effort goes into overcoming thread friction, about 42% goes into overcoming friction between the rotating member face and the piece being clamped, while the remainder goes into stretching the bolt.

Looking at the action of this clamping force, one sees that it is absolutely critical to effective performance. A good example is a bus or truck wheel. The wheel bolts' function is to tightly clamp the wheel against the hub so the friction between the two members can drive the wheel and, if the bolts can't provide enough clamping force, they see a fatigue load and rapidly fail.

We will discuss bolting practices later but a wheel hub on a car is a good example of the need for intelligent tightening procedures. If the lug bolts aren't tightened adequately, they will fail from fatigue and the wheel will fall off. On the other hand, if they are tightened nonuniformly they will cause distortion of the wheel hub and warp the brake disk or drum.

Three important points to understand are:

1. The clamping force must be greater than the separating force if fastener fatigue is to be avoided.

2. On a gasketed assembly uniform clamping is critical to minimize warpage and distortion and give reliable performance.
3. Over tightening can cause not only component distortion but also thread seizure.

Pursuing this, about 25 years ago the Bolting Technology Council (this was a technical group that has since been absorbed by and whose work has been continued by the ASTM) hired Dr. Andre Bazergui of L'Ecole Polytechnique of Montreal to analyze those things that could affect torque vs. tightness in a bolt. Dr. Bazergui did an excellent job with a detailed study looking at an exhaustive variety of possible conditions and modifiers and reported that the most important properties, in order of severity, are:

1. Proper installation (install the bolt squarely)
2. Thread lubricants and plating
3. Materials
4. Installation practices
5. Everything else, i.e., corrosion, surface smoothness, tightening method, number of threads engaged, hand or power tightened, etc.

From this study, we can see the importance of consistent bolting practices and we will return to that later in the chapter.

BOLT STANDARDS

There are a numerous groups around the world that have standards concerning bolts and, combining the different bolt strength levels and materials, there are hundreds of possible material combinations. In turn, those bolts are made by thousands of different companies and one of the questions a failure analyst frequently hears is "How do you know how strong the bolt is or who made it?" The answer can be found by looking at the bolt head and the markings.

According to U.S. law and the Fastener Act of 1991 and ISO Standards, unhardened low carbon steels, the weakest of the steel bolting materials, don't require any head markings. However, all heat-treated general-purpose bolts and all strain hardened bolts and studs must have the grade and manufacturer's identification clearly marked on their heads and Table 12.1 shows a list of the common U.S. threaded bolt specifications and their identification symbols. In addition, ASTM standards require the manufacturer's identification and grade marking on all ASTM A 307 bolts.

How do you find out who made that bolt? Fastener Technology International, a technical magazine published by an Akron, Ohio, organization, maintains a registry of manufacturers and their symbols. To determine who made a fastener you can buy a copy of their roster of trademarks and search. Some are easily recognized abbreviations like RBW (for Russell, Burdsall, & Ward)

TABLE 12.1

Common Inch-Series Fastener Specifications

Grade Designation (ASTM & SAE)	Material and Condition	Tensile Strength (psi, min)	Proof Load (psi, min)	Size Range (in)	Head Identification
General					
A 307 Gr A	Low/medium carbon steel not heat treated	60	na	1/4–4	ASTM requires trademark
A 307 Gr B	same	60 min 100 max	na	1/4–4	ASTM requires trademark
SAE Gr 1	same	60 min	36	1/4–1.5	Not required may have trademark
A 325	Low/medium carbon steel, Q & T	120	85	1/2–1	A 325
		105	74	1.1–1.5	Trademark req'd
SAE Gr. 5	Medium carbon steel, Q & T	120	85	1/4–1.5	
A 354 Grade BC	Medium carbon alloy steel, Q & T	125	105	1/4–2.5	BC
		115	95	2.6–4	Trademark req'd
A 354 Grade BD	Medium carbon alloy steel, Q & T	150	120	1/4–2.5	BD
		140	105	2.6–4	Trademark req'd
A 449	Medium carbon steel, Q & T	120	85	1/4–1	A 449
		105	74	1.1–1.5	Trademark req'd
		90	55	1.6–3	
SAE Gr. 8	Medium carbon alloy steel, Q & T	150 min 170 max	120	1/4–1.5	
A 490 Type 1	Medium carbon steel, Q & T	150 min 170 max	120	1/2–1.5	A 490 Trademark req'd
A 490 Type 2	Low carbon alloy steel, Q & T	150 min 170 max	120	1/2–1	A 490 Trademark req'd
Low Temperature					
A 320 (tested @-101°C)	Medium carbon alloy steel, Q & T	125	105	up to 2.5	L7 Trademark req'd
A 320 (tested @-73)	Medium carbon alloy steel, Q&T	100	80	up to 2.5	L7M Trademark req'd
High Temperature					
A 437	Martensitic stainless, Q&T	145	105	All	B4B
A 437	Martensitic stainless, Q&T	115	85	All	B4C
A 437	Medium carbon alloy steel, Q&T	125	105	up to 2.5	B4D

while others are relatively hard to interpret symbols but the registry contains almost all the major suppliers from around the world. (The U.S. patent office also maintains a registry of makers marks.)

General purpose metric bolts can easily be identified because they have their strength markings on the heads, always using a decimal point between the tensile strength value and the proof test multiplier. Photo 12.1 shows a metric bolt head and Table 12.2 shows a chart of the common metric fastener designations along with the comparable SAE and ASTM standards. The first number on the head shows the tensile strength in hundreds of megapascals while the number after the "." is the proof test multiplier. In the photo, the 8.8 shows a bolt where the tensile strength is approximately 800 megapascals (MPa), the proof test multiplier is 0.8, and the resulting proof strength is approximately 640 MPa. The SBE on the bolt head is the manufacturer's trademark and, looking in the *Fastener Technology International* database, the manufacturer is Societa Bulloneria Europa SPA from Monfalcone in northeastern Italy.

We should be very careful to note that all U.S. thread steel socket-head cap screws, frequently called Allen head cap screws, are quenched and tempered medium carbon alloy steels with a minimum tensile strength of 150,000 psi (1040 MPa). This is the specification for an SAE grade 8 bolt but it only applies to steel U.S. thread socket-head cap screws. It doesn't apply to stainless steel or other alloys and metric socket-head cap screws can be found in all seven classes listed in Table 12.2.

(Adding further confusion, we have found fasteners made outside the United States with steel socket head capscrews with inch-dimension threads that were heat treated to the equivalent of an SAE Grade 5 bolt.)

PHOTO 12.1 Both the manufacturer and the strength of the steel in this metric bolt are clearly stated on the bolt head with the 8.8 the tensile strength and yield strength multiplier and the SBE the manufacturer's authorized mark.

TABLE 12.2

Common Metric Fastener Specifications[4]

ISO Class[1] Identification (ASTM F 568)	Size Range	Material[2]	Tensile Strength (MPa)	Yield Strength (MPa)[3]	Approximate SAE/ASTM Equivalents
4.6	M5–M100	Low or medium carbon steel	400	240	SAE Gr. 1 A 307 Gr A
4.8	M1.6–M16	Annealed low or medium carbon steel	420	340	SAE Gr. 1
5.8	M5–M24	Low or medium carbon steel, cold worked	520	420	SAE Gr. 2
8.8	M16–M72	Medium carbon steel, Q & T	830	660	SAE Gr. 5 A 449/A 325
9.8	M1.6–M16	Medium carbon or low carbon martensitic steel Q & T	900	720	
10.9	M5–M100	Medium carbon steel, Q & T	1040	940	SAE Gr. 8 A 354 BD A 490
12.9	M1.6–M100	Alloy steel, Q & T	1220	1100	A 574

Notes:

[1] Property class is the required identification on the fastener head or the stud body. All fasteners 5.8 and stronger must also be marked with the manufacturer's trademark.

[2] Q & T = quenched and tempered

[3] Proof strength as determined by the yield strength method.

[4] SAE metric specifications are covered by Standard J1199

[5] Some metric bolts and studs may be marked with ASTM designations such as A 325M. The addition of the M indicates a metric fastener that meets the requirements of the ASTM standard.

Nut Standards

Just as there is a variety of bolt strengths, there is also a variety of nut strengths however, because of the way the stress is applied, nuts are inherently stronger than the bolts and studs they are attached to. Visualizing a cross-section of the assembly shown in Figure 12.1, the shank of the bolt and the bolt threads are in tension, while the nut body and threads are in compression.

It is extremely uncommon to see a properly matched nut fail. They can't fail from fatigue, because the threads are in compression, and the actual mechanism is almost always an overload failure with dilation from overstress, followed by the threads stripping. There are numerous steel nut standards but a good rule of thumb to follow is shown in Table 12.3.

TABLE 12.3

Guidelines for Selecting the Proper Steel Nut Strength

Approximate Bolt or Stud Tensile Strength (metric/US)	Typical Bolt Grades	Required Nut Material
400 to 500 MPa/60,000 to 80,000 psi	M – 4.6, 4.8, 5.8 US – A 307, SAE Gr. 1 & 2	Any nut steel can be used but unhardened is preferred
725 to 825 MPa/105,000 to 125,000 psi	M – 8.8, 9.8 US – A325 & 449 SAE Gr. 5	Any nut steel can be used
1000 MPa/150,000 psi	M – 10.9 US – A 490, SAE Gr. 8	Nut should be made from steel that is comparable in strength with the bolt or stud

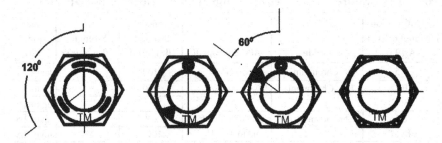

FIGURE 12.2   This shows two groups of U.S. nuts with their strength markings. There are two other marking systems, but they all follow the same basic concept. Note that the two nuts on the left side have markings 120° apart, similar to the markings on a grade 5 bolt, and these nuts are the equivalent material. Then the two on the right, with markings 60° apart, are compatible with Grade 8 bolts. Nuts marked with a 2H are designed for B7 bolting.

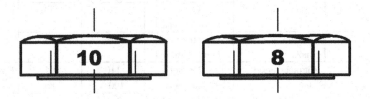

FIGURE 12.3   Two metric nuts and standards call for the tensile strength grade to be stamped on a wrenching flat. This shows the nut on the left has a tensile strength of 1000 MPa while the one on the right is 800 MPa. Unlike the metric bolts, there is no requirement for a yield strength multiplier. Like U.S. nuts, those with no grade on them are the weakest.

What is confusing about nuts is how they are identified. Metric nuts, as with metric bolts, are marked with the strength class but U.S. thread nuts have several different systems as seen in Figures 12.2 and 12.3.

MATERIALS

Almost everything we've written has been about steel bolting materials but there are a huge variety of other bolting materials available. The two charts above actually list only 10 different materials but the current ASTM section on fastener standards cover more than 500 pages. Table 12.4 shows some common ASTM standards for high temperature alloy and stainless bolts and studs. In addition, there are other ASTM standards such as F 593 that covers general purpose stainless fasteners and F 468 that covers nonferrous fasteners made of copper, nickel, aluminum, titanium and their alloys. (F 468 alone specifies 28 different alloys.)

One of the important points to realize with Table 12.4 is the huge number of alloy combinations in standards such as ASTM F593 and even small differences, such as the underline beneath the *B8*, *B8M* and *F 593F* can be extremely important. In the

TABLE 12.4
High Temperature Inch-Series Alloy and Stainless Fastener Specifications

Identification Symbol (ASTM A 193)	Grade	AISI Equivalent	Tensile[1] Strength (ksi, min)	Yield[1] Strength (ksi, min)	Size range (in)
B5	5% Chromium	Type 501	100	80	up to 4
B6,	12% Chromium	Type 410	110	85	up to 4
B6X			90	70	up to 4
B7,	Chrome–moly	4140, 41, 42, 45	125	105	up to 2.5
B7M		up to 4145H	100	80	up to 2.5
B16	Chrome–moly–	none	125	105	up to 2.5
	vanadium		110	95	2.5 to 4
B8	Chrome–nickel	Type 304 Solution treated	70	30	all
B8	Chrome–nickel	Type 304–sol'n	125	100	Up to 3/4
		treated and	115	80	3/4 to 1
		strain hardened	105	65	1. 1–1.25
B8C	Chrome–nickel	Type 347 Solution treated	70	30	all
B8M,	Chrome–nickel– moly	Type 316 Solution treated	80	35	all
B8M	Chrome–nickel– moly	Type 316–sol'n	110	95	Up to 3/4
		treated and	100	80	3/4 to 1
		strain hardened	95	65	1. 1–1.25
F 593A	Chrome–nickel	Type 303, 304,	65–85	20	1/4–1.5
F 593C		etc.	100–150	60	1/4–5/8
F 593F	Chrome–	Type 316, 316L	75–100	30	1/4–1.5
F 593F	nickel–moly		110–150	75	3/4–1
F 593H			85–140	45	3/4–1.5

[1] In some cases a strength range is noted.

last four lines of the figure are *only five of the 33 different alloy combinations* listed in the F593 standard, and standard F 594 covers the nuts and nut marking. (In U.S. standards, the nuts are marked either on the top of the nut or on the wrenching flats.)

So, there are not only a huge number of bolting materials but also a large number of plating treatments that can be used on those bolts. The reason for repeatedly mentioning this is that changing bolting materials has two significant effects:

1. As bolting materials get stronger, they become less able to absorb impact loads and more susceptible to corrosion damage and embrittlement.
2. Different materials and coatings have differing friction characteristics and variations in the friction change the amount of torque that goes into stretching the bolt.

If you can't get a direct replacement material and have to change material specifications in a bolting application either really know what you are doing or ask a professional for a recommendation in writing. Why do we say to get a recommendation in writing? Like everything else, specifications on nuts and bolts look fairly simple and straightforward from a distance. Up close, where you have to understand the effects of fatigue and impact loading and the possibility of corrosion and embrittlement failures, the job is much more complicated. By putting the recommendations in writing "the expert" will carefully review what is actually happening.

FATIGUE DESIGN

Peterson's Stress Concentration Factors edited by Pilkey, is a book that shows examples designed to reduce stress concentrations. One of the illustrations is of a bolt where the body is turned down, reducing the stressed area but, because of the reduction in the stress concentration factor, the reduced diameter bolt as shown in Figure 12.4 is actually stronger in fatigue than the original.

How much can the bolt body be reduced? Any reduction in the body diameter below the minor diameter of the threads reduces the static load carrying ability of the bolt. In most cases this isn't as important as the dynamic (fatigue) load capacity and Peterson's book covers these in detail. It is interesting to see that the bolt body diameter can usually be reduced significantly below the thread root diameter before the fatigue strength is reduced. However, there are several points that should be well recognized when reducing that diameter. They are:

1. When a bolt is turned down, careful attention has to be paid to the radius under the head, the radius at the threads, and the surface finish of the body. The radius under the head should be at least as large as the original radius (It can actually be larger because of the increased diametrical clearance

in the washer and hole.) and must be smooth and well blended. The radius leading from the body to the threads should be about the same as the head radius. Leaving tool or chatter marks and/or surface irregularities perpendicular to the axis of the bolt will introduce additional stress concentrations. The machinist has to do a quality job.

2. Most bolts under 29 mm (~1 1/8″) have rolled threads. These rolled threads are actually stronger than cut threads in fatigue applications because they have residual compressive stresses. Cutting into the bolt body can remove the compression forces but if it is machined as outlined above, there should be no net reduction in strength.

The comment on thread rolling above brings up the subject of bolt manufacturing and Figure 12.5 shows how, starting with a piece of wire, a bolt is commonly manufactured. Some larger bolts are machined from bar stock, but just about everything bought over the counter is cold forged with rolled threads.

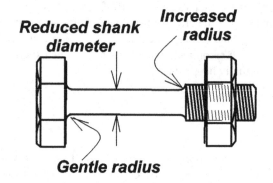

FIGURE 12.4   A bolt designed for fatigue applications.

Seven Steps to Forming a Bolt

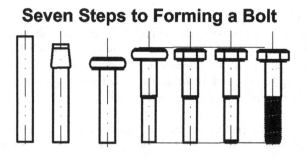

FIGURE 12.5   A common seven step procedure for forming bolts.

ASSEMBLY PRACTICES

Using the patented device shown in Figure 12.6, we have conducted hundreds of analyses on a variety of fasteners, reviewing some of Dr. Bazergui's findings. One of these tests showed that lubricant on threads can reduce the typical torque needed to develop a given clamping force by about 40%. Another series of tests showed that lubrication results in much more uniform tightening, i.e., there was far less scatter in the torque vs. tightness data for the lubricated bolts. A third series of tests showed that reusing bolts that have not been distorted or visibly damaged rarely has a significant effect on the clamping force. Based on these tests (and other data) we recommend lubricating all nuts and bolts, unless the equipment manufacturer specifically recommends otherwise.

This difference between lubricated and unlubricated bolts is especially obvious when stainless steel or aluminum bolts are used. Tests on the demonstrator show that typically an unlubricated stainless or aluminum bolt will only develop about half the clamping force of a lubricated one before the bolt seizes to the nut. After the bolt seizes (galls) it usually has to be cut apart because the two pieces have become welded together and disassembly is impossible.

BOLTING PATTERNS AND SEQUENCES

There are lots of bolting practices and written procedures but one that we should start off with involves the basic properties of the bolting material. Table 12.5 shows the allowable elastic deformation of the three common bolting property groups. If bolts are tightened so the elongation is more than the listed values, plastic deformation of the bolt body occurs and, because the force is directly related to the physical metallurgical properties of the bolt material, that gives a very closely controlled value. This *torque-to-yield* practice is often used in the erection of steel structures and bolting of cylinder heads on engines because it assures that the clamping force is uniform. (But one thing that should be emphasized is that the bolts still have to be tightened in steps and in a well-designed sequence!)

Almost every critical assembly where tightening of multiple bolts is involved requires a procedure specifying the steps in which the bolts should be tightened

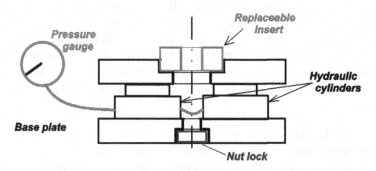

FIGURE 12.6 Our device for studying the torque-tension relationship for various bolting practices.

TABLE 12.5

Common Bolt Ranges and the Allowable Elastic Elongation

Approximate Tensile Strength (metric/US)	Typical Grades	Allowable Stretch (mm per mm of grip) (in per in of grip)
400 to 500 MPa/60,000 to 80,000 psi	M–4.6, 4.8, 5.8 US–A 307, SAE Gr. 1 & 2	0.002
725 to 825 MPa/105,000 to 125,000 psi	M–8.8, 9.8 US–A325 & 449 SAE Gr. 5	0.004
1000 MPa/150,000 psi	M–10.9 US–A 490, SAE Gr. 8	0.006

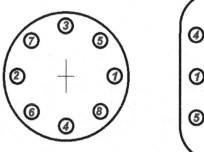
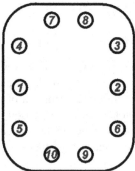

FIGURE 12.7   Two common bolting patterns.

and a sequence such as shown in Figure 12.7. The goal of these bolting patterns is to reduce distortion of the bolted members and result in more uniform clamping force and more reliable assembly life. But it is also important point to be certain that the stresses on the joint are consistent with the design and don't include serious misalignment forces. For good reliability the flanges have to be properly aligned before the bolts are assembled and tightened. This means you shouldn't use a come-along to pull the flanges into alignment and the flange faces should be parallel. Current industry practices (and ASME Boiler and Pressure Vessel codes) call for:

- Flange flatness – Must be within 0.15 mm (0.006″). The maximum allowable out-of-flatness shall not occur in less than 20° of arc.
- Flange alignment – The two mating flanges must be parallel to within 1 mm in 200 mm (1/16th in per foot) and the centerlines of the two flanges must be within 3 mm (1/8 in) maximum offset

The value of these standards is that they improve long-term flange and piping system reliability.

There are many different tightening sequences and for important assemblies we generally recommend the following:

1. Inspect the fasteners to ensure all meet the same specifications.
2. Liberally lubricate the bolts, both on the threads and under the side that will be rotating.
3. Lightly snug the bolts to ensure good alignment and seating.
4. Tighten the bolts in a sequence to 60% of the maximum recommended torque, then 80%, and then 100%.
5. Repeat the sequence at 100% of the rated torque.
6. Go around it twice clockwise tightening to the specified torque.

This bolting practice has been in common chemical and petrochemical plant use for over 50 years and, as far as we can determine, it was developed empirically and has produced very good results.

The only formal study we know of where clamping forces were accurately measured on a gasketed multi flange steel piping assembly was conducted by the University of Dayton for the ASTM. In that outstanding experiment, a pair of 18″ flanges and a wafer valve were bolted together to try to understand the actual effect that tightening one bolt has on the clamping force of the others. The bolts were equipped with strain gauges that read out on a computer screen, then tightened to prescribed clamping forces. As each bolt was tightened the changes in the stress within the others could easily be seen. In the initial trial, the bolts were tightened in three steps, lightly snug, 80% of rated clamping force, and 100% of clamping force. When the 100% round was completed, they found that the ratio of clamping forces between the tightest and loosest bolts was 9:1! Following another round of tightening to 100%, still in the cross pattern, produced a ratio of 5.5:1. Then two successive circular paths resulted in a ratio of about 2.5:1. We suggest the six steps shown above because in industrial practice it is typical that torque wrenches are used, not strain gauges, and the ability of torque to measure clamping force is much poorer than that of the strain gages.

What is truly important to realize is that nonuniform tightening will bend flanges and will result in leaks. The difficulty in recognizing this is that the leaks often don't show up until sometime later when the gasket can't cope with the errors, deformation, and residual stresses caused by the poor tightening practices. A leaking flange is a failure just as a broken bolt or a failed bearing and our studies found the most common causes of leaky flanges involved improper tightening practices by people who don't understand how to do the job correctly.

A story that should go a long way toward illustrating the importance of careful tightening procedures happened many years ago at the plant where I worked. Early one morning I was about 25 m (80 ft) off the ground on top of a vessel, measuring the flows into it as part of a reliability project. We had a well-intentioned but hard of learning mechanic who was give the assignment of replacing a 30 cm (12″) cast iron pipe elbow. The mechanic walked up and began attaching the cast iron elbow to some steel piping. From far above and a long distance away, while measuring the flows, I watched him rig the elbow into position, put a bolt on one side at the 3:00 o'clock position and then put another bolt at the 9:00 o'clock position. He then picked up an impact wrench and first tightened one of the bolts, then the other. As he was tightening the second bolt the whole piping system jumped as the flange broke. It was horrendously noisy, there was a long way between us, and there was little I could do.

About half an hour later the mechanic returned with another cast iron elbow. He rigged it into position, put a bolt in one side, and put a bolt in the other side. He then picked up his impact wrench and tightened first one bolt and then the other. *Again*, as he was tightening the second bolt, the piping system jumped as the flange broke off the elbow. I shook my head, silently wondered how some people even survived, and went back to measuring flows.

Another half later he came back with another 12″ short radius cast iron elbow. When I saw him returning, I decided that was enough flow data and started to climb down to talk with him!

We hear stories like this and there are several common reactions:

1. We have all worked with people like these and we chuckle or laugh about the things they've done.
2. We all know that when the cast iron flanges broke, there was probably enough stress to bend the mating steel flanges "just a little".
3. We all know enough not to bolt up a cast iron flange like he did. But we all regularly see examples of people using impact wrenches to tighten flange bolts, one after the other, going around in a circle, and making no pretense of trying to use good bolting practices.

Until we think about it, few of us recognize that there will be problems and probable leaks the next time that flange is taken apart and rebolted. Without looking at the flange flatness, we might blame the people who just put the flange together for using poor practices.

Before leaving the subject of bolting practices we should review why thread lubrication is an important point. Earlier in the chapter, we said, "We recommend lubricating all nuts and bolts, unless the equipment manufacturer specifically recommends otherwise". Sometimes, as with automotive wheel studs, the manufacturers state that no lubricant should be used. In other cases they may specify the lubricant.

We strongly recommend following the manufacturer's recommendations but in other applications, those where the manufacturers don't state specifically how to lubricate a threaded fastener (and this probably includes 99% of the bolts used in the maintenance world), the specific lubricant is not anywhere near as important as being consistent in the bolting procedure. In the first few pages of this chapter we mentioned that about 82% of the torque goes into overcoming friction while 18% goes into stretching the bolt. If that coefficient of friction is increased by 10%, (e.g., from 0.120 to 0.132) that means 90% of the torque would go into overcoming friction and the tightening force would be reduced by more than 40%. In the design of a bolted joint such as a steam or process line flange, the factor of safety on the bolt strength is almost always a very, very conservative number. What invariably causes the problem isn't that the bolts are too weak, it is that they aren't tightened uniformly. Being consistent with a lubricant eliminates one of the possible errors.

There are almost an infinite number of combinations of base metals, plating, and lubricants and unless there are specific instructions to the contrary, for the best reliability we recommend standardizing on an assembly lubricant and applying it liberally.

FASTENER FAILURES

Fasteners fail mechanically either from overload or from fatigue. With overloads the bolt fails immediately as the load is applied. The actual fracture may be ductile, with lots of distortion, brittle with no visible distortion, or somewhere in between. Photo 12.2 shows the ductile elongation of a tie bolt from the stator of a large motor and if it had been tightened a little more the bolt would have broken in a classic cup and cone ductile elongation failure.

Overload failures usually happen as the bolt is installed and most are the result of over tightening but they can occur at any time if the bolt is subjected to severe loads. Water hammer, dropped parts, and thermal shock loads are all examples of situations where an installed, well-designed and successfully operating bolt can fail due to overload.

Depending on the material the failure usually looks like the bolt above or a classic brittle fracture with chevron marks pointing to the origin. However, there are

PHOTO 12.2   This is a bolt from a wheel motor on a 240+ton haul truck. To ensure consistent clamping force on the motor components, the bolt yield strength has been slightly exceeded.

also failures where the fastener is sheared, i.e., literally cut in half. These overload failures can be recognized by their deformation in the direction of the shear force.

FATIGUE FAILURES

Fatigue is by far the greatest cause of fastener failures and happens when the clamping force developed by a bolt is less than the separating force applied to it. The concept that a bolt has to be tightened to the point where the developed clamping forces are more than the separating forces is sometimes difficult to understand because one of the fears people have is that a little more load on the bolt will then result in fracture. However, although the spring constant of the clamped materials must be taken into effect, from a practical standpoint, the bolt only sees a measurable change in stress when the separating force exceeds the clamping force.

Two solid verifications of the value of tightening until the clamping force is greater than the separating force can be seen in the following well-documented projects. In one case an ultrasonic flaw detector was used to routinely monitoring a group of 28 large cyclically loaded bolts. Over the first year of the project we found several cracks and developed the ability to predict their growth. Then we found another bolt, one with a very small crack. It didn't have to be changed right away and maintenance was backlogged so we continued to run the machine for several months while the crack slowly grew larger. Eventually a hydraulic torque wrench was used and tightened the bolt to the required 17,500 ft-lbs. After this retightening the machine ran for an additional 30 million cycles and several inspections found no additional crack growth before the bolt was changed.

The second example involved a pump shaft assembly where the impeller was held on by a stud and nut. Over many trials we found that tightening the nut to 25% of the rated torque produced an assembly that would consistently fail in about 30,000,000 stress cycles. Tightening to 50% resulted in a life of about 100,000,000 cycles, while tightening to the full design value resulted in an assembly that never failed.

These and other examples are essentially the same and effectively prove how important it is to develop a clamping force greater than the separating force if fatigue failures are to be avoided.

In inspecting bolt failures one of the clues to inadequate clamping force is that there is movement between the bolts and the mating pieces. The two examples shown in Photos 12.3 and 12.4 have both failed from fatigue and the photos show areas where the evidence of movement is relatively obvious.

Photo 12.3 is of a body-fitted bolt. Looking at the shank, it can be seen that the half closer to the nut still has the original machining marks on it while the section on the right has been polished smooth. This polishing is proof that there was movement between the bolt and the pieces that were supposed to be tightly clamped together. In many similar examples, especially with electro-galvanized bolts, the side of the bolt body will show an irregular blackened and discolored fretted area.

In a similar manner, the piece shown in Photo 12.4 is a nut and a failed bolt. The fracture was the result of a fatigue crack starting at about the 1:30 o'clock position. The contact surface on the underside of the head shows two polished areas, one at about 10:30 o'clock and one at 2:00 o'clock. The only way these surfaces could have

PHOTO 12.3   The fretting on the right half of this bolt body proves that it was loose.

PHOTO 12.4   A fatigue failure, but the interesting point is the irregular contact on the nut surface.

become polished during operation is if the mating surface was irregular and there was movement between the two pieces.

One of the contributors to the failure in Photo 12.4 was the irregular contact surface and the importance of a well lubricated, smooth contact cannot be overstated. After a bolt is tightened, over time it will normally lose a portion of its clamping force because of minute smoothing of the asperities and plastic deformation. If the contact area is small the forces can be extremely high and there may be substantial plastic deformation.

The Grade 8 bolt shown in Photo 12.5 is an extreme case of this but does a good job of illustrating the point. It was used on a hydraulic pump and fractured after being in operation for several months. The photo clearly shows a fatigue failure with several ratchet marks and several prominent progression marks. Inspection of the bolt body shows it was bent by about 0.6 mm (0.025 in) and there is visible fretting on the shank just below the origins. Even though the bolt was tightened, the load was concentrated at one point at the head. The bolt flexed during peak load spikes and

PHOTO 12.5   It's unusual that the head of a bolt will fail, but this one was slightly bent. The progression marks on the fracture surface clearly show where it started and the fracture plane describes how the effective force changed as the crack progressed.

fatigue cracking resulted. (This bolt has been electroless nickel plated to preserve the surface features.)

HYDROGEN EMBRITTLEMENT AND HYDROGEN INFLUENCED CRACKING

I'm a firm believer that, at least when it comes to materials, there is no such thing as a free lunch. The much-too-common plant solution for a bolt failure is to replace the failed bolt with a stronger one, but there is always a trade-off and higher strength bolts are much more affected by corrosion and hydrogen influenced cracking (HIC).

(The terminology used to describe hydrogen damage leads to confusion. The most common term used by fastener manufacturers is *hydrogen embrittlement* and they talk about failures due to hydrogen absorbed during processing or use that then causes cracking. In the eyes of corrosion engineers and scientists this type of cracking is HIC. They reserve the term hydrogen embrittlement for those instances where the hydrogen causes a change in the chemistry of the metal, e.g., when hydrogen turns the metal into a metal hydride.)

In those situations where a higher strength bolt is used, they usually cost more, almost always are more susceptible to corrosion and the resultant fractures, and typically have less impact absorbing ability than the lower strength bolts. There is general agreement among fastener experts that bolts harder than HRC 38 shouldn't be used in *any corrosive applications* because of their tendency toward HIC. In the chemical industry there are numerous prohibitions against using bolts harder than HRC 22 in applications involving exposure to acids because of possible HIC. Grade 8 bolts can be harder than HRC 38 and most Grade 5 bolts and A 325 bolts will be harder than HRC 25. (We have seen hydrogen cracking of bolts that were HRC 27.)

From a metallurgical standpoint, the problem is that the absorbed hydrogen causes a stress concentration within the metal at a microscopic crack tip. The mechanisms of hydrogen embrittlement and hydrogen influenced cracking are discussed in Chapter 7 and identification of them requires detailed metallurgical analysis.

FAILURE LOCATIONS AND BOLT DESIGNS

Figure 12.8 shows the common failure locations and is the result of a large study done by an Italian researcher. The failures occur in these areas because of stress concentrations and the causes and the failure locations haven't changed much over the years.

The concept of *stress concentrations* was discussed in Chapter 5 and earlier in this chapter. Because they cause the local stresses in a part to increase, it's a concept that everyone involved with fasteners and fastener problems should thoroughly understand. They are caused by changes in metallurgy, changes in shape, shrink fits, and material defects and are sometimes called *stress risers*. The consequence of them is that, in the immediate area of the stress concentration, the actual stress is much greater than the average stress in the part. The technical symbol for the stress concentration factor is K_t and the factors for various thread shapes usually fall in the range of 4 (for well-rounded shapes) to 8 (for acme and stub acme threads).

GENERAL COMMENTS AND CAUTIONS ON BOLTING

- *Welding on fasteners –*
  - Unhardened bolts can be safely welded with no significant metallurgical changes.
  - Welding on heat treated bolts, such as Grade 5 and higher, will change the metallurgy of the bolt. Is the new bolt stronger or weaker? Is there a stress concentration at the end of the rehardened zone? How serious are the residual stresses? Has hydrogen been introduced? These are some of the metallurgical questions that have to be answered if a welded heat treated bolt is going to function reliably.
  - In addition, what affect has the heating had on the bolt preload? Has it reduced the preload so that an assembly that started out with adequate clamping force now sees significant fatigue stresses? In tests on the bolt demonstrator, several times a bolt was tightened to the rated preload, then lightly tack welded on the head. The heat and resultant plastic elongation resulted in permanent preload reduction of about 50%.
- *Reuse of nuts and bolts* – I wouldn't reuse a nut and bolt assembly on a critical job, but a reused bolt can effectively be as safe as a new one if you

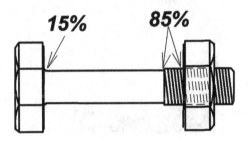

FIGURE 12.8 This shows the locations where fatigue cracking occurs.

check to see that there is no visible rust or corrosion and the threads are not distorted. How should you check for thread distortion? – Run a nut down the threads with your fingers and it shouldn't bind. (Sometimes the cost of inspection is more than the bolt cost but if a bolt isn't distorted or corroded our opinion is that it can be reused.)

FAILURE EXAMPLES

EXAMPLE 12.E1

This is a U.S. thread, 3.7 cm (1.5″), Grade 8 low alloy steel bolt and there is rust on it.

It was used to anchor a large press platen and operated in a high humidity atmosphere at about 25°C (77°F). There are several of these bolts and they can be impact loaded if a piece is dropped onto the platen. The plant has had a history of similar failures over the past few years and they are in a location where an inspection is difficult.

Inspection – A close examination of the fracture origin, centered at about 7:00 o'clock, shows some small corroded and discolored pits. The fracture surface is uniform in smoothness up to the shear lip that runs from about 3:00 o'clock to 9:00 o'clock. (Shear lips are common on tension failures and occur when the crack changes direction as it grows across the piece. This change in direction happens because the load center is no longer in the physical center of the piece.) Shadowing the fracture surface, it can be seen that there are a series of chevron marks that point toward a relatively broad origin. Examination of the contact surface of the head showed no polishing, fretting, or other evidence of movement.

Conclusions and recommendations – This is a brittle overload failure and happened as the load was applied. We recommended that a metallurgical examination of the bolt be conducted before any other action. This would determine if the fracture was influenced by hydrogen damage. Whether it was or not, look at the operating practices to better understand the impact loading and why it occurs. If corrosion influenced the failure a specially coated bolt should be considered (Photo 12.E1).

PHOTO 12.E1

PHOTO 12.E2

Example **12.E2**

This is a nut from a 1/2″ B8 stainless steel bolt, part of a SS bolt assembly that is used at ambient temperatures in a damp atmosphere and it failed suddenly during assembly.

Inspection – The fracture face is almost flat and perpendicular to the major axis. The face shows the swirl marks and eccentric eye that are typically seen in a ductile overload torsional failure. In addition, the threads in the nut are deformed and there was no evidence of lubrication.

Conclusions and recommendations – It appears to be a ductile overload failure that occurred during tightening when the threads seized. (If there is a question about whether it occurred during assembly or removal, a metallurgical examination can be made. That will show a deformed layer along the surface and the direction of the deformation will answer the question.) We recommended that future bolts be lubricated with a good grade of EP lubricant (Photo 12.E2).

Example **12.E3**

This is one of a group of 35 mm (~11/2″) diameter fine thread bolts used to hold the upper frame to the base of a massive press in a forging shop. The material meets the specifications for a B7 bolt and the previous bolt in this location had broken and been replaced. The plant is located in the northeastern United States and the operating atmosphere varied from generally being close to outdoor ambient on weekends to being 10–20°C (18–36°F) warmer during the two shifts of the operating week.

Inspection – The fracture face shows a fatigue failure starting at about the 8:30 o'clock position in Photo 12.E3A. At the origin there are several progression marks, then there is another progression mark about 1/4 of the way across the face, with the IZ occurring at the 3/4 mark. The entire fracture face is essentially perpendicular to the long axis of the bolt and, as shown in Photo 12.E3B, the tips of the right hand threads have been worn smooth. There is a great deal of grit and carbon dust on the entire bolt.

PHOTO 12.E3   (A and B)

Conclusions and recommendations – The progression marks and change in surface roughness show that the bolt failed from fatigue. There are several ratchet marks but the change in surface color shows they occurred after the cracking had started. (The ratchet marks indicate a high stress concentration and are common with a threaded fastener.) The surface shows several distinct changes in roughness with definite progression marks. The rounded tips of the threads show the base plate has been rubbing against the bolt and the press pieces were moving side to side relative to each other. Based on these, our opinion is that the failure was from low cycle fatigue caused by occasional peak stresses from very heavy forging loads. The recommendation was that they remove, inspect, lubricate, and retighten the remaining bolts using several steps in the procedure. In addition, if additional failures occur it may be that the mating surfaces are no longer flat to each other and careful shimming is needed.

Example 12.E4

Photo 12.E4 shows cracking in the underhead radius of one of a group of 610 mm (~2.5″) bolts that anchor the lower die set of a 500 ton forging press. The press runs in a heated and controlled atmosphere and does not see the environmental variations

PHOTO 12.E4

of the previous example. There are eight bolts that hold the die set in position and we had been asked to nondestructively test several sets of them. They had been using a hydraulic torque wrench to tighten the bolts.

Inspection – A magnetic particle inspection found cracks in about 35% of the inspected bolts and we then looked carefully at the machining. There were different radii under the bolt heads and it appeared that the bolts had been machined by different shops from different materials. Some of the underhead radii were very generous while those on the cracked bolts were small and machined in obvious steps. In addition, hardness testing of the bolt materials showed that the uncracked ones were about HRC 30 while the cracked ones were HRC 20. After we looked at and found the cracks, one of the maintenance personnel found a broken bolt that had been left in a corner of the building and that fracture was obviously from fatigue.

Conclusions and recommendations – The bolts failed from fatigue indicating that either they had not been tightened correctly or the machine frame was flexing. However, the replacement bolts they were using had been made from a sample rather than a drawing and:

1. The underhead stress concentration was higher than it should be.
2. The material had only 66% of the fatigue strength that it should have had (HRC 30 vs. HRC 20).

We recommended they change the material specification back to the original and ensure any replacement bolts were machined to the same specifications as the originals. We also recommended they go around the die set bolts in sequence at least twice in the tightening procedure (Photo 12.E4).

Example 12.E5

This is a bolt from a hydroelectric plant runner coupling on a shaft that ran at about 300 rpm. (The "runner" in a hydroelectric turbine is the propeller-like device that is turned by the water and the coupling is a rigid coupling between two shafts.) This 40 mm (~1.5″) body fitted bolt assembly met the requirements of ASTM A 193 B7 and had been in essentially continuous operation for eight years when it fractured, i.e., over 1.2 billion cycles.

Inspection – The fracture face shows the classical fatigue failure characteristics. The failure started at the very bottom of Photo 12.E5A where there are several ratchet marks. The fracture face was flat and the crack gradually grew across the body, getting progressively rougher and rougher until the final fracture occurred about 70% across the shank. Looking at the underside of the bolt found polishing (from the 10:00 o'clock to the 11:30 position in the photo) where there had been movement between the bolt and the coupling face and the machining marks were worn off. Then, looking at the side of the bolt in Photo 12.E5B, some fretting can be seen immediately under the bolt head.

Conclusions and recommendations – Based on the observations, the bolt had been loose in the coupling assembly for a time, the head contact was not good and:

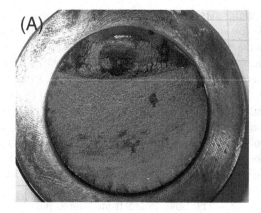 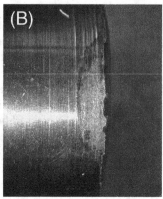

PHOTO 12.E5 (A and B)

- The fatigue crack started where the surface had fretted. (A common origin.)
- The IZ indicates the bolt was moderately stressed.

Confusing was that, with those conditions we might expect a similar bolt to last 20 or 30 million stress cycles, but certainly not 1.2 billion!

The failure occurred in late April and discussions with the plant personnel found they had had serious ice problems about a month before the failure. We also interviewed the crew that had done the installation eight years earlier and discussed their bolting procedures. Based on the evidence, our opinion was that the likely cause was ice falling onto the assembly during the previous month.

We also suggested that, when the parts were reassembled to:

1. Check to be sure the fit of the coupling assembly is secure.
2. Check the bolt and bolt hole perpendicularity.
3. Lubricate the bolt threads and the contact face under the nut with an anti-seize compound and tighten the two coupling bolts to 3400 nm (~2500 ft-lbs) in four steps, then recheck the final torque reading.

Example 12.E6

This is an igniter from a large gas engine and is similar to a spark plug used in a typical automotive engine. The site had many of these engines and when an igniter failed they had to be replaced. However, frequently during attempts at removing the old igniter, the threads would seize and gall, requiring lengthy and expensive repairs. The problem had not been seen with the OEM igniters but these were a less expensive replacement and not quite the same construction.

Inspection – The inspection found the threads were essentially being welded into position and this galling is not uncommon on assemblies that see severe temperature cycling. Hardness testing of the steel male and female thread assemblies found they were of the same hardness.

PHOTO 12.E6

Conclusions and recommendations – The igniters were galling because of the compatibility of the materials when subjected to heat and pressure. We suggested they could go back to the OEM igniters or they could speak with the new supplier and ask that they change to a harder threaded body and consider a thread coating that reduced the probability of welding to steel (Photo 12.E6).

EXAMPLE 12.E7

These are two bolts that were used to support a hoist in a manufacturing area. The facility was light manufacturing in the New York–Baltimore industrial complex and the room where the hoist was used was generally between 10°C (50°F) and 35°C (95°F). The hoist was used near a hydrochloric acid system and there were occasional traces of acid fumes in the air. The hoist was cycled about 100 times per month and the weight of the lifted load was less than 5% of the bolt's static capacity.

Inspection – The two hanger bolts were aligned above the hoist as shown in the photos with the hoist support midway between them. Examining the bolt in the right photo shows that it failed from fatigue with the crack starting at the top and growing downward, across the hoist axis. In addition, there was a substantial amount of rust on the mounting block, the piece shown in the background. The left bolt also showed a fatigue face with a crack that worked its way about one-third of the way across the bolt before the catastrophic failure occurred. This bolt was much more heavily stressed than the first one and the lack of corrosion on the mounting block shows this fracture happened immediately before the hoist fell.

Conclusions and recommendations – From the direction of the crack growth on the first bolt it is apparent that the crane was side loaded. The first bolt failed from these excessive loads and then the second bolt was carrying the full load and subjected to a bending moment, and it rapidly failed from fatigue. (The real problem was that work area operations were changed, but the equipment wasn't. This

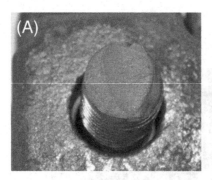 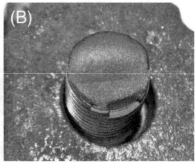

PHOTO 12.E7   (A and B)

failure happened because of obvious human [process engineering] and latent errors [Photo 12.E7].)

EXAMPLE 12.E8

This is a 50 mm (~2″) Class 10.9 bolt from the counterweight of a large piece of mining equipment. There are several of these bolts that hold the weight in position and the outboard ones periodically fail. Loss of the counterweight could be very dangerous because the machine could tip instead of lifting.

Inspection – As can be seen in Photo 12.E8A, a straightedge placed along the side of the bolt shows that it is bent off to one side. Then, looking at the opposite side it can be seen that the edge of the fracture face has been deformed outward. Photo 12.E8B shows that the fracture surface is very flat and perpendicular to the body, but it isn't uniform in smoothness and doesn't have any signs of a fatigue failure's instantaneous zone. A close examination shows the surface features are all distorted in the same direction as the overhang shown in photo A.

The bolts showed no evidence of looseness.

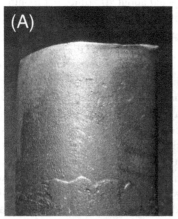 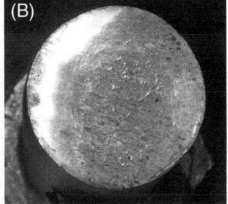

PHOTO 12.E8   (A and B)

Conclusions and recommendations – The fact that there is deformation shows this is either a VLC fatigue or an overload failure. There is no distortion in the direction of the long axis of the bolt so it wasn't an axial load that caused the failure, it was a shear load. The smoothness of the fracture surface is indicative of a bolt that experiences a tremendous side force, essentially cutting the bolt in two. This was not a fatigue failure so the forces that caused it were applied immediately before the failure occurred.

We suspect that these counterweight bolts were being impact loaded by unusual strains on the machine. We did not get to see the specific machine where the failures were occurring but did watch similar machines and they had the potential for occasional shock loads. We recommended they:

- Talk with the factory design personnel to see what their FEA shows would actually happen during heavy loading.
- Keep a log of failures including who was operating the machine and what the machine was being asked to do.
- Add the bolts to the pre-shift inspection procedure.

Example 12.E9

Photo 12.E9 shows the lug nut off a car wheel. We had ordered a new set of tires and wheels and had installed them on our car, tightening them in three steps, at 50, 75, and 100% of the rated torque. (Following the manufacturer's recommendations, there were no thread lubricants used.) We had driven the car about 12 miles when an odd noise developed from the left front tire. Stopping the car, we found the lug nuts on that wheel were loose and we retightened them using just a tire iron. After we got to our destination, I used a torque wrench to tighten those nuts to the correct reading, and then found the nuts on the other three wheels were also loose! We drove the car home and, again, the nuts were loose and retightened and we made sure that car was only used for *very* short trips, and the nuts were rechecked after each trip. After this happened a couple of times, I pulled one of the lug nuts off and examined both it and the wheel with a microscope.

Inspection – Examining the aluminum wheel found an erratic contact pattern and lots of plastic deformation. Then we removed a group of the lug nuts and checked the concentricity of the tapered fits with the centerlines of the threads and found the runout was as much as 0.05 mm (~0.002″).

Conclusion – The runout of the nut meant that the wheel was riding on only a small part of the taper that would eventually plastically deform, and the nuts would work loose. We contacted the national tire and wheel vendor and they immediately sent us another set of lug nuts. We installed those nuts and have had no more loose wheel problems.

Example E12.10

This isn't a failure analysis as much as a glaring example of what not to do! The photo shows a hoist at the end of a jib crane and a careful examination of the area above the nut shows the nut has been welded to the eye bolt.

PHOTO 12.E9

What effect has the weld had on the metallurgy of the eyebolt? We don't know! But what we do know is that welding on bolts is something that should only be done after a careful engineered analysis.

Welding on a tightened bolt will stress relieve it and hasten a fatigue failure. Welding on bolts with a significant carbon content, over about 0.25% can cause cracking.

One of the first examples of welding on a bolt that I was asked to analyze involved a 64 mm (~2.5″) diameter crane hook shank where an overzealous repairman had welded the end of the nut. The resultant change in the metallurgy caused the hook to brake in half as they were picking up a one-ton spreader bar that they were going to use to pick the rotor out of a large steam turbine (Photo 12.E10)!

Example 12.E11

The plant had a huge press that used eight 760 mm (~30″) OD cylinders and they had broken the tie rod on one of them, so we were asked to ultrasonically test the other 82 mm (~3.25″) tie rods used on the press hydraulic cylinders. Photo 12.E11 shows the fracture face and, from looking at it, we can tell that the failure was the result of fatigue cycling. The failure occurred at the transition where the threads met the smooth body of the bar.

Inspection – The fracture face shows a large fatigue zone and an instantaneous zone that is approximately 25% of the total area. Also, the fatigue zone is extremely smooth and this indicates that the fracture is relatively slow growing.

There were 27 of these tie bars that hadn't failed, and the ultrasonic flaw testing found cracks in five of them. Our recommendation to the plant was, that, to eliminate the fatigue cycling they should tighten the bolts to 115% of what the manufacturer specified. The manufacturer's representative was there and he refused to let them

PHOTO 12.E10

PHOTO 12.E11

tighten the bolts any more than the original specs, so they did that and found most of them were loose to some degree. 50,000 stress cycles later we retested the bolts and found that the growth in four of the five cracked rods had stopped. One had continued growing and we again suggested they tighten the nuts to 115%, and they did that.

Conclusion – Testing after the second retightening found that all of the cracks had stopped growing, once again proving that, if the bolt is tightened so the preload is greater than the operating load, the bolt can't fail in fatigue.

BIBLIOGRAPHY

ASM Handbook, Volume 1 (8th edition), Edited by T. Lyman, ASM International, Materials Park, 1967.

Avallone, Eugene, Baumeister, III, Theodore, and Sadegh, Ali M., *MARKS Standard Handbook for Mechanical Engineers*, McGraw-Hill Education, New York, 2006.

Louthan, Jr., M.R., Hydrogen Embrittlement of Metals: A Primer for the Failure Analyst, Savannah River National Laboratory, Aiken, SC, 2008, http://sti.srs.gov/fulltext/WSRC-STI-2008-00062.pdf.

Pilkey, Walter D., *Peterson's Stress Concentration Factors* (2nd edition), Wiley-Interscience, Hoboken, NJ, 1997.

Sachs, Neville W., *Failure Analysis Made Simple: Shafts and Fasteners*, Reliabilityweb.com, Ft. Myers, FL, 2018, ISBN: 978-1-941872-81-9.

Speidel, M.O., "Corrosion Fatigue in Fe-Ni-Cr Alloys," in *Proceedings of the International Conference on Stress Corrosion and Hydrogen Embrittlement of Iron-Based Alloys*. Unieux-Firminy, France: National Association of Corrosion Engineers (NACE), 1977.

13 Miscellaneous Machine Component Failures – Chains, Lip Seals, Couplings, Universal Joints, and Plain Bearings

CHAINS

In understanding chain failures, we should start by pointing out that most U.S. chains are built to ASME standards that were written many years ago. The standards describe the physical properties of the chain but do little to describe the quality of it. There are manufacturers that build chain as a commodity item to meet the minimum specifications and there are other chain manufacturers whose products offer much, much more than meeting a minimum standard.

There is a huge variety of roller chains, silent chains, engineered chains and custom chains and, despite the different names, they share many properties and a general failure analysis procedure can be applied to all of them. Three factors to keep in mind are the materials, their relative action, and their lubrication. Also, a misconception to recognize is that you will often hear people talk about chain "stretch", but what actually happens 99.999% of the time is that the chain components wear: that causes the pitch to increase slightly, and people think that the chain has stretched.

There are British Standard chains that are very similar to ANSI chains and have Imperial (inch) dimensions. But there are some differences between the two and the British chains should be used on British Standard sprockets. There are also metric chains.

CHAIN DESIGN

Most chains used in industry are roller chains. They are similar to a bicycle chain and the components are shown in Figure 13.1. The component contact, the movement of bushing on the barrel and the barrel on the pin, is all rolling or sliding and lubrication is critical. To improve the wear life, most of the contacting components are case hardened, although some pins are through hardened.

There are some lightweight chains that eliminate the barrel but the wear mechanisms are all similar.

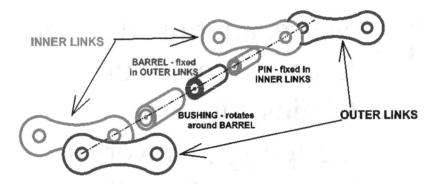

FIGURE 13.1   The components of a typical industrial roller chain. (There are light duty chains that eliminate the barrel.)

Chain drives designed "by the book" are typically designed to last from 15,000 to 20,000 operating hours before the chain needs replacement and, because of all the relative movement and contact, lubrication has a tremendous effect on chain life. In addition, other important influences result from:

- Sprocket sizes – The more teeth on the sprocket, the less the chain flexes and the less wear.
- Slack – Excessive slack will result in more chain movement and more wear.
- Misalignment can concentrate the forces on one side of the chain instead of having the two sets of side links working together.
- Sprocket wear – As the sprocket wears the pitch changes and that creates additional stress that reduces the chain life.
- Corrosion – The chain components are hardened and, with the effect of hydrogen cracking on hardened steels, corrosion can cause chain breakage.

WEAR

Regardless of how much the chain is lubricated, it is eventually going to wear and a gage can be used to check the relative condition of the chain. In the active part of the chain set up a measure over some convenient distance, such as two chain pins 300 mm (~12″) apart, then periodically remeasure the distance between the pins. When the elongation exceeds about 2.5% the wear rate will increase dramatically because the chain has worn through the hardened case.

The chain wears every time it flexes and a chain with a lot of slack and frequent turns, as shown in Photo 13.1, will wear out much more rapidly than one with only a slight sag.

Sprocket wear is important because of the effect on the chain stress. As the sprocket wears the pitch changes and, if a new chain is run on worn sprockets this will concentrate the entire load on the last link or two as opposed to distributing it over several links. The increased load will increase the wear rate and reduce the chain life. One way to try to counteract this is to harden the sprockets and our experience is that, in most applications, hardening the sprockets to about HRC 45 or more

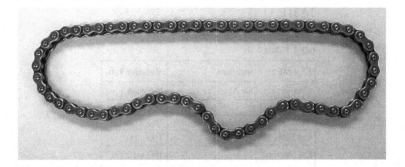

PHOTO 13.1   This roller chain has been taken off the sprockets, but arranged in a manner so the "waves" in the bottom of the photo duplicate its appearance during operation. The chain was loose and at every change in direction additional wear took place.

will improve the system life when compared to the normal surface hardened HRC 35 sprockets.

How much sprocket wear is tolerable? That really depends on the nature of installation (the chain size, the sprocket size, and the application) and the cost of the chain. On small chains (bicycle-sized), or small sprockets, almost any visible wear will cause skipping of the chain. In most industrial roller chain applications, 3 mm (1/8″) wear is excessive.

> The chain and sprocket wear on our bicycles gives an interesting insight into this question. We have a tandem bicycle and do a lot of touring with it. Because it uses standard chain and sprocket components, the drive to the rear wheel wears fairly quickly, much, much quicker than on either of our individual bikes. Changing the rear cluster, the gears on the back wheel, is about four times as expensive as changing the chain and it is very time consuming. As a result, we change the chain when the wear is about 1% and have seen a good increase in sprocket life. One other point that can be seen from bicycle applications is an appreciation for minimum number of tooth specifications. As the rear sprockets become smaller the wear rate increases dramatically.

Lubrication

The chain components that rub against each other are commonly case hardened to around HRC 50. As the chain flexes during operation these pieces wear and eventually the hard and abrasive wear particles form a lapping compound. Lubrication provides two functions:

1. With lightly lubricated components, where lubricants are applied by an occasional brush or drip application, the lubricant helps to reduce the wear.
2. With spray or dip application the lubricant actually washes out the wear particles.

TABLE 13.1

The Effect of Operating Conditions on Chain Life

Operating Condition	Relative Life
Corrosive and/or Abrasive	less than 1
Clean and Dry	1
Manual (intermittent) lubrication	10
Oil Bath	100–500
Oil Stream	1000–10000

A study by Rexnord, Inc. showed the following effects of lubrication (Table 13.1):

The lubricant used should be an EP oil and if the chain is being used in a wet atmosphere it should also have additives for improved corrosion resistance. A general guideline for the lubricant viscosity is that it should be between 60 and 120 cSt at the operating temperature.

CHAIN BREAKAGE

We have never seen a chain break where there haven't been obvious and excessive loads.

Photo 13.2 is of a link from a size 160 chain and looking at it several interesting points are immediately apparent:

1. The end of the pin shows that it is case hardened with an obvious case around the entire exterior.
2. The fracture face on the end of the pin shows a fatigue failure that starts at about the 12:30 position and also shows the pin was heavily loaded in bending.
3. The irregular wear on the side of this outer link shows there was substantial misalignment.

Some guidelines on chain breakage are:

1. Pin failure
   a. Look at the direction of the crack growth. Generally caused by worn or badly aligned sprockets.
   b. Could be an impact load that starts a fatigue crack that takes a long time to grow across the pin. When this happens the IZ is usually very small.
2. Roller
   a. Failure – Most likely a result of a badly worn sprocket where the teeth hit the rollers, but this can occur with a badly worn chain on a high-speed drive. (The sprocket in Photo 13.3 was off a dirt bike where the owner had put a new chain on only five hours before the sprocket and chain were removed. In that time, a third of the rollers had been broken off the chain.)

PHOTO 13.2   The fatigue failure of a case hardened chain pin. The origin is at the top and there are obvious progression marks heading toward the relatively large IZ. This failure was the result of gross misalignment.

PHOTO 13.3   This badly worn sprocket came off a dirt bike and, because the wear changed the effective center distance of the teeth, destroyed the chain in less than five hours.

    3. Bushing failure – Usually the result of prolonged system overloads.
    4. Side link failures – Look at the failure face.
      a. Instantaneous fracture is almost always a single overload.
      b. Fatigue failure is usually the result of prolonged system overloads.
      c. HIC is from corrosion acting on the hardened steel.

PHOTO 13.4 Look at the wear on the sides of the teeth on these two sprockets from an industrial drive that was horribly misaligned. This drive ran for only two weeks and eventually broke the chain.

5. Cotter pin failures are essentially always the result of misalignment. (Cotters are designed as keepers to hold the pieces in position and are not supposed to be subjected to substantial thrust [side] loads.) (See Photo 13.4.)

A suggested failure analysis procedure for a chain drive is:

- Find out how long it was in run and the actual lubrication practices.
- Compare the actual loading with the design loading. (How do they compare with the recommended chain capacity?)
- Layout and reassemble the failed pieces.
- Identify the failed components.
- Inspect, preferably using low power magnification, the links and other components to see if they were bent, deformed, or worn.
- If pieces were broken, was it fatigue, overload, or hydrogen influenced cracking (HIC)?
- Inspect and describe the wear pattern on the pins, rollers, and side plates.
- How well were the sprockets aligned?

LIP SEALS

The life of the current design of lip seals is generally in the range of 8,000 to 10,000 hours. Beyond this they can still provide a measure of sealing but can't prevent leakage. There are dynamic sealing devices offered by several manufacturers. They are much more expensive than a lip seal and usually require special machining of the housing but they last and protect the machinery essentially indefinitely.

The conventional lip seal design is such that the lip rides against the shaft and the shaft surface finish is critical. Figure 13.2 shows a sketch of a shaft and the running surface specification from several major seal manufacturers. Of interest is that for

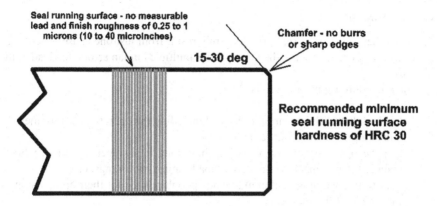

Seal running surface - no measurable
lead and finish roughness of 0.25 to 1
microns (10 to 40 microinches)

Chamfer - no burrs
or sharp edges

15-30 deg

Recommended minimum
seal running surface
hardness of HRC 30

FIGURE 13.2 Typical recommended surface conditions for effective shaft and seal life.

optimum life they also recommend the shaft be hardened to at least HRC 30 because the seal lip tends to wear it away.

In any installation where there is rubbing contact lubrication is critical and:

- A well-lubricated seal lip will outlast a dry one by thousands of hours.
- At the typical surface speeds seen in industrial machinery multiple lip seals are usually shorter lived than the single lip seals because of the heat generated at the unlubricated second lip.

This second category really involves Arrhenius' Rule and the exponential increase in degradation rates as temperatures increase.

A challenge faced in all lip seal applications is that of the lip being flexible enough to meet the imperfect operating conditions, i.e., it can't have high runout and high speed and high STBM tolerances and a decent seal life. (See Figure 13.3.) For reasonable life, high misalignment means the application has to accept either much lower speed or much less runout. On the other hand, if it is a high-speed application it can't have high runout and high STBM.

We have used aftermarket stainless steel sleeve inserts to repair shafts with worn lip seal grooves with good success in both industrial and automotive repairs. (The work hardened stainless steel has good surface finish roughness control.)

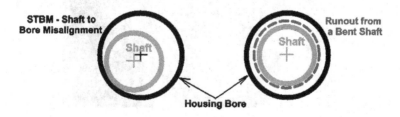

STBM - Shaft to
Bore Misalignment

Shaft

Runout from
a Bent Shaft

Shaft

Housing Bore

FIGURE 13.3 Two common sources of seal problems.

FAILURE ANALYSIS

As with other equipment, seal failures rarely result from just one cause. There may be a single physical cause but there are always multiple human errors that lead to the failures.

The procedure we usually use is:

1. Look at the conditions around the seal including the operating environment and equipment temperature ranges.
2. Get the operating history including how long it has been installed, who installed it, any machine data about load ranges and swings, etc.
3. Talk to the operators and maintenance people asking for their opinions on the source of the problem.
4. Using low power magnification, preferably with a binocular microscope, look at the seal components, the lip, the housing, and the shaft surface.
5. From the above, diagnose what happened.

The seal problems we've seen usually begin at installation when the lip is damaged or the seal shape is distorted. Photo 13.5 shows the lip of a seal used on a piece of medical equipment. The manufacturer was plagued by user complaints about oil and grease leakage from the equipment and was convinced that the seal manufacturer had a problem with their materials. It wasn't until they examined the seal lip using a 10X magnifying glass that they saw the small nick on the contact path just at the horizontal centerline of the photo, and realized it was caused by their installation techniques.

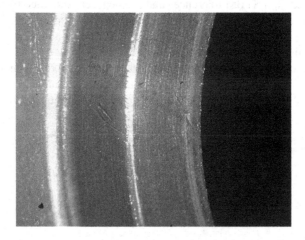

PHOTO 13.5 In the center right of the photo, on the light gray running surface of the seal, can be seen a small diagonal nick that was caused by careless assembly procedures.

FLEXIBLE COUPLINGS

I think about flexible couplings and consider them to be a little like a necessary evil. If we didn't have them, we would destroy bearings and break shafts, but we do have them and they have all sorts of idiosyncrasies. You have to be careful about lubricating the ones with steel grids and gears and the elastomer ones tend to have diabolical torsional characteristics that sometimes destroy other components. Worst of all is that uneducated people don't realize that:

1. They are expecting this device to run for billions of revolutions. (An 1800 rpm machine makes 946,000,000 revolutions in a year.)
2. You pay for careless alignment with higher vibration resulting in shorter bearing and seal lives and increased energy usage,

GRID AND GEAR COUPLINGS

Gear and grid couplings are designed to accept a certain amount of misalignment but the factory suggested maximum allowable misalignment has decreased significantly in the past 30 years. It isn't that the couplings have changed, but that the damaging effects of misalignment have become better recognized.

One of the challenges with these couplings is that they tend to centrifuge the grease used to lubricate them and, with most greases, the thickener is denser than the oil/additive combination and it's the thickener that ends up at the OD where the lubrication is needed.

At low speeds, below about 200 rpm, the effect of the centrifugal force isn't a problem and almost any NLGI 1 or 2 AW/EP grease can be used, but as speeds get higher the use of a specialty coupling grease is mandatory to get reasonable life.

With that coupling grease and proper alignment, the relubrication interval can be as follows:

- Grid couplings – every five years
- Gear couplings – every three years

However, when these couplings are relubricated, it's necessary to open and physically clean out the exhausted grease before repacking with fresh grease. (We have seen people use a fitting on one side of the coupling to pump grease in until it comes out of the port 180° away and think they've relubricated the coupling. At our request, when they have opened the coupling up, they realize that the grease has just made a path between the hubs and never got to the critical areas that are still packed with exhausted grease.)

One of the interesting things about inspecting failed couplings is that they are relatively outspoken about how they failed. For example, looking at Photo 13.6, one can see:

1. From the wear, the grid imprints on the inside of the cover, it is apparent that the hubs were seriously misaligned.

PHOTO 13.6 This coupling is a disaster in several respects as the text mentions. One of the interesting clues to the skill and direction of the workforce is that there was no sign of the rubber gaskets that should be on the hubs.

2. From the wear on the ends of the male grid driving pieces, the rounded surfaces on the right hub, it can be seen that the grid was tight up against them and the center gap spacing was incorrect.
3. There was rubbing against the hubs on both sides and the two paths from the seal retainer can be seen, but the seals were not installed.
4. Obviously it had been a long time since it was lubricated.

From this, we know that the person misaligned it and didn't properly space it on installation. They also didn't put the seals in. From this flagrant series of errors and the lack of knowledge of the importance of alignment procedures, we would also suspect that their management has problems.

In contrast to the disaster mentioned above, look at the coupling grid in Photo 13.7. This was in operation on a skip hoist for about nine months and appears to be in excellent condition but still tells an interesting story as follows:

1. The wear patterns are a little heavier on one side than the other, suggesting that the drive forces may not be quite balanced in both directions.
2. For nine months of operation the amount of wear is very small and the coupling is well lubricated.
3. The polishing on the top of the grid shows it is rubbing against the cover and the coupling is misaligned.

The gear coupling hub shown in Photo 13.8 is badly worn and inspection of the teeth around the hub shows that they are not uniformly worn, proof that the alignment wasn't good. The coupling has been in run for about two years and was lubricated with "a good grade of grease". The combination of misalignment and improper lubricant has resulted in the premature death of another coupling.

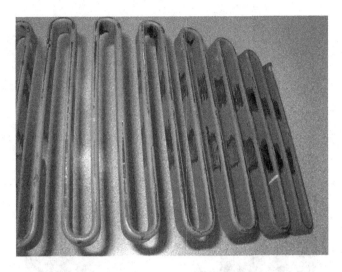

PHOTO 13.7   This is what a coupling grid should look like, except look at the polishing on the top of the grids, the result of misalignment. (Using a specialty coupling grease, we once ran a grid coupling for six years at 1200 rpm and it looked about like this one when we finally removed it!)

PHOTO 13.8   This ran for two years at 1775 rpm and wasn't well aligned. In addition, the grease they used was a conventional NLGI #2 and not a coupling grease, so the thickener, and not the oil and additives, was centrifuged to the area of the teeth.

Photo 13.9 shows three months of operation on the hub from a 400 hp, 3550 rpm motor that drove a compressor until the drive end motor bearing failed. The hub shows substantial wear on the teeth and, like the example immediately above, the wear is not uniform around the hub, indicating there was a serious misalignment. (There was angular misalignment to go along with the 0.4 mm [~0.015″] parallel

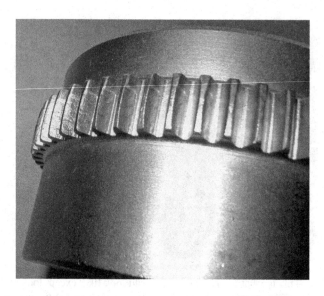

PHOTO 13.9 The "killer coupling" that, in only four months at 3600 rpm caused the death of the motor and the compressor – because it wasn't properly aligned. The wear on the teeth is a combination of poor lubrication and misalignment.

misalignment.) This was in a small waste treatment plant and the crew that installed the assembly didn't understand the alignment requirements and didn't have any grease – so they looked around until they "found something" to put in it. The drive end bearing on the compressor failed about a month later!

DISK COUPLINGS

Most have the machines with disc couplings that we've worked on have been critical to the processes and the maintenance personnel have taken great pains to be certain that the running alignment has been correct.

We have seen the occasional coupling with disks fatigue fractured like the ones in Photo 13.10. Those couplings have usually seen some misalignment, although we worked on one in a nuclear power plant where there was a tremendous torsional resonance that visibly flexed the disc pack.

ELASTOMER COUPLINGS

One of the hard-to-believe points about elastomeric couplings is that some of the elastomers deteriorate on the shelf more rapidly than in operation. We had a Rex Omega coupling as a class demonstrator for several years. One day someone dropped it and the elastomer element broke in two.

Fortunately, that type of failure doesn't happen very often and the two major problems we've seen with elastomer couplings come from torsional vibration and misalignment.

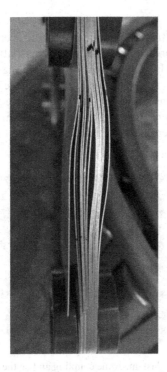

PHOTO 13.10 This shows some broken disks, the result of repeated fatigue forces from misalignment, although they could also be caused by torsional vibration in the system.

Torsional vibration is relatively common with elastomer couplings and one of my frustrations is that, despite seeing it many times, I don't have any good photos to use in this book. There two common symptoms of torsional vibration as a problem are:

1. With couplings such as a T.B. Woods that use a "toothed" elastomer insert that fits into the two hubs, there is wear on both sides of the teeth.
2. With couplings that have an elastomeric element that is bonded to the hubs, such as the Rexnord Omega mentioned earlier, the elastomer develops a series of interlocking diagonal cracks

In both of these cases the damage is the result of the coupling flexing back and forth as the shaft rotates, i.e., it is not rotating smoothly.

If you suspect torsional vibration, one way to check for it is to use a strobe light and freeze one coupling hub. If the other one is twitching back and forth or appears blurred, there is a torsional vibration and the only way to cure it is to change the torsional critical frequency of the system.

The shaft example shown in Photos EX 6.24 and 6.25 was from a large fan that ran for years at about 880 rpm. Then they bought a VFD drive and found the airflow at about 720 RPM was ideal for their operation. Unfortunately, it was also ideal for a torsional resonance and the shaft obviously failed from torsional fatigue. We also got to look at the coupling and it had elastomeric bushings that were absolutely melted from the cyclical loading. VFD drives can be a problem because every system has some frequency where there is a torsional critical frequency and if the machine is running close to that speed the forces can be greater than what the design called for.

PHOTO 13.11   The orange dust under the coupling and on the motor vents has been ground off the teeth of the polymer insert by serious misalignment.

Misalignment is another common problem because too many mechanics feel that, because they have flexible elements, there isn't the need for good alignment. Photo 13.11 is of a 75 kW (100 hp) water pump and the orange dust beneath the coupling has been worn off the teeth of the T.B. Woods coupling insert. (In operation, you can tell if the damage is from misalignment or torsional vibration by looking at the coupling with a strobe light. If the two hubs can be frozen at the same time, the problem is from misalignment.)

With couplings like the T.B. Woods', the deflection from any misalignment is taken up by flexing of the insert's center section. The force on the teeth of the insert is supposed to be enough to prevent movement between the hub and insert however, if the misalignment is excessive the teeth will rub and move against the hub and eventually wear off.

Also, if the coupling is oversized and there is misalignment, the force on the teeth won't be enough to prevent movement and the insert will rapidly wear out.

UNIVERSAL JOINTS

There are several types of universal joints but the only industrial ones I've worked with have been the classical Cardan joints, also known as Hooke's joints, such as those found on a car or truck driveshaft. Most of the failures have been of the caged needle bearings in the cups and, whenever a U-joint fails inspection of those bearings is valuable. (There is a U-joint cup in the kit we travel with that came off of

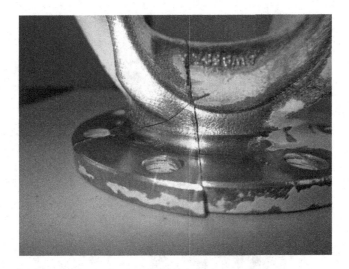

PHOTO 13.12   A fast growing fatigue crack in a steel Cardan joint yoke.

a paper machine and always drove in one direction, but there are significant false brinell patterns at 0° and 180°, proof that there was a torsional vibration and that the actual loads were at least double the design loads.)

Photo 13.12 is of a yoke from a universal joint and the shaft it was on had two Cardan joints to cancel out the changes in angular velocity. Inspection of the photo shows a cut line starting at the bottom center and extending up and slightly to the left. There is a second line, the original crack, that starts on the far left side and runs diagonally up to the right. The joint was cut apart so we could look at the loose piece to the left of it to understand the type and age of the failure. As you can guess from the fracture angle, it resulted from torsional fatigue and the surface indicated that it was relatively fast growing.

Photo 13.13 in another failure but this one is a classic case of an instantaneous brittle overload, with some tremendous chevron marks that point to the failure starting at journal at the top of the photo.

PLAIN BEARINGS (JOURNAL BEARINGS)

Plain bearing lubrication has been around since the invention of the wheel and all sorts of oils and greases have been used. Some of the most intelligent early research was conducted by Beauchamp Tower, and a sketch of the device he used to prove how they worked is shown in Figure 8.16. For convenience, Figure 8.3 is duplicated here and shows how the lubricant film is developed in the bearing. Also, it bears repeating that, if enough oil is fed into the bearing, the lubricant film will support the load. But as the load increases, the lubricant temperature also increases and that results in a lower viscosity oil that escapes more readily.

Some plain bearings are made from aluminum alloys and some are made from bronze alloys, but most are made with a variety of babbitt metals facing the shaft.

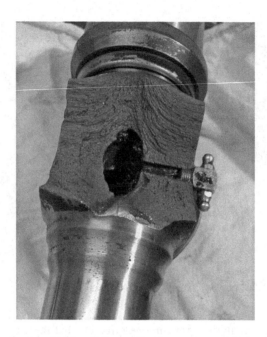

PHOTO 13.13 Another U-joint failure off of a paper machine, but this one was instantaneous.

The original babbitt bearings had a high lead content but, with the environmental and health regulations, they have largely been replaced by tin-antimony-copper alloys.

One of the great features of babbitt metal is its low melting temperature and its "embeddability". If there are solid particle, larger than the film thickness, in the oil fed to the bearing, the particles will rub against the rotating shaft, get hot, and slowly embed themselves into the low-temperature melting babbitt (Figure 13.4).

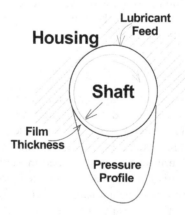

FIGURE 13.4 A view of a plain bearing showing how the lubricant film is developed. It is the viscosity of the oil and the rotation of the shaft that determine the oil film thickness.

I once made the mistake of putting bronze bearings in a machine because I felt they would be stronger and last longer than the existing babbitt bearings. We didn't do a good job of filtering the oil and some particles larger than the oil film eventually got in the between the bearing and the 10,000 rpm shaft. Instead of embedding themselves in the bearing material, like they would have if it had been babbitt, they jammed between the shaft and the bronze scoring the shaft, and that set up a serious vibration, and that destroyed the oil film, and the shaft rubbed through the bearing, and the friction of the steel shaft on the cast iron bearing housing lit off the lube oil, and that pretty much burned up the rest of the 300 kW (400hp) machine. It cost us several thousand dollars per operating second and if we had used babbitted bearings it probably wouldn't have happened! Sometimes we learn the hard way.

Most plain bearings in applications like car and truck engines will eventually fail from wear when those tiny particles eventually erode the layer of babbitt, increasing the clearances and allowing greater leakage. However, in industrial applications most of the plain bearings failures that we have seen have resulted from fatigue and much of the text and photos discussed here will address them.

Plain (journal) bearing construction – Most plain bearings, such as those in our cars and trucks, are "trimetal bearings" and consist of a steel backing, with thin layers of copper and babbitt metal on top of them. Almost all the others are babbitt metal on top of heavy steel or cast iron supports.

Babbitt metal is relatively weak and has a low fatigue strength so both the thickness of the layer and the quality of the bond between it and the backing metal is critical, and research has shown that, the thicker the babbitt, the more likely that a fatigue crack will occur.

We have seen an immense number of failures that resulted from poor adhesion and have also been told by babbitt bearing manufacturers, "If the fatigue forces on the bearing are high enough, all babbitt bearing will eventually fail from fatigue cracking". As shown in Figure 13.5, the way these fatigue failures happen is that:

- The top layer of babbitt is microscopically flexed until a crack starts.
- The crack grows through the babbitt layer.

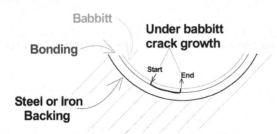

FIGURE 13.5 The pattern of crack growth in a babbitt bearing.

- It then grows along intermetallic layer at the bond line.
- Then returns to the surface and the unsupported piece cracks loose like an iceberg.

(We have seen several examples of large babbitt bearing fatigue cracking that, hard to believe, didn't result in catastrophic bearing failure or shaft damage. These happened on some large engines where we removed the connecting rods because of high babbitt content in the oil and found the bearings had large areas, maybe 20% of the surface area, that had cracked into loose pieces. The pieces were floating up against each other like pieces of sea ice and, if we had continued to run the engines there would have been a disastrous failure because the pieces rubbing against each other resulted in areas with no babbitt at all. But the interesting thing to us was that, even with the floating pieces, the surface developed an oil film and the bearings were still doing their job.)

Below are a series of babbitted bearing failures, and the photo captions describe the failure mechanism and key points (Photos 13.14–13.21).

PHOTO 13.14   This is the upper half of a ring-oiled motor bearing. Note how, in what should have been the unloaded side of the bearing, the babbitt has cracked out because the bond was defective.

PHOTO 13.15 This is the lower half of a fan bearing and the upper is barely visible at the bottom. Note how the babbitt has cracked out of the side of the bearing, yet the load zone is on the bottom. This is poor bonding between the babbitt and backing.

PHOTO 13.16 This is a trimetal insert from a gasoline engine. The center of the thin babbitt layer has worn through, showing the crankshaft journal running against it is barrel shaped.

PHOTO 13.17   This shows another ring-oiled motor bearing with the top half on the left. Interesting is that the pattern shows gross misalignment with contact on only one edge of the bottom. The cause was excessive belt tension on the machine drive.

PHOTO 13.18   This is a pair of inserts from the main bearings of a 3000 kW (4000 hp) gas engine and the wear pattern shows serious misalignment. The engine was loose on the foundation and the looseness allowed the crankcase to distort.

PHOTO 13.19   This is the bearing from a steam turbine and the sandblasted appearance is the result of static discharge from the dry steam. It was controlled by putting a grounding ring on the shaft. (The ring needs replacement about every five years.) We've seen similar damage on large motors.

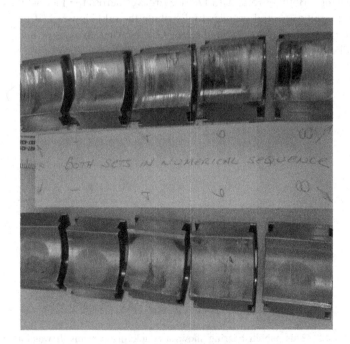

PHOTO 13.20   Two sets of tilting pad bearing inserts that show electrical discharge damage. As with the bearings discussed earlier, there are portions of the bearings that look as though they were sandblasted. These and the ones above were detected by oil analysis and the machines shut down before more serious damage occurred.

PHOTO 13.21   An insert out of an industrial compressor and the large areas cracked out on only one side clearly show this is an example of poor bonding in the manufacturing process. The black dots are patches of babbitt that have been heated and oxidized.

The photo below is of a very different failure cause. Plain bearings often have grooves or recesses to allow the oil or grease to be distributed across the entire face of the bearing. (Both 13.15 and 13.17 have grooves across the face while 13.18 and 13.21 have circumferential grooves to distribute the oil.) However, if those grooves are located in the high pressure load zone of the bearing, they allow the lubricant to escape (Photo 13.22).

PHOTO 13.22   This babbitt bearing shows a couple of problems. It was one of a series of failures but the damage on this was less than the others and provides a better example. There's a problem with misalignment, but equally serious are the grooves in the load zones that effectively allow the lubricant to escape to lower pressure areas and reduce the size of the bearing's load carrying area.

BIBLIOGRAPHY

Bearing Failure Analysis, Glacier Bearings, London, UK, 1960.

Chain Installation and Maintenance, U.S. Tsubaki, Wheeling, IL, 1992.

Elwell, Richard C., Foreign Object Damage in Journal Bearings, *Lubrication Engineering*, Vol. 34, No. 4, 1978, 187–192.

Fuller, Dudley D., *Theory and Practice of Lubrication for Engineers* (2nd edition), John Wiley & Sons, Inc., 1984, ISBN: 0-471-04703-1.

A General Guide to the Principles of Operation and Troubleshooting of Hydrodynamic Bearings, Kingsbury, Inc., Philadelphia, PA, 2008.

Hoffmeister, W.F., *Lubrication of Drive and Conveyor Chains*, Rexnord Inc.

Klamann, Dieter, *Lubricants and Related Products*, Verlag Chemie, 1984, ISBN: 0-89673-177-0.

Lubes 'n' Greases, numerous articles, LNG Publishing Company, Inc., Falls Church, VA.

Maintenance Suggestions, Diamond Chain Company, Indianapolis, IN, 1981.

Sachs, Neville W., *Failure Analysis Made Simple: Bearings and Gears*, Reliabilityweb.com, Ft. Myers, FL, 2015, ISBN: 978-1-941872-30-7.

Tribology and Lubrication Technology – the monthly journal of the Society of Tribologists and Lubrication Engineers, Park Ridge, IL (numerous articles).

PPFA – A Glossary of Technical Terms

Chevron marks: Arrow-like indications on the surface of a brittle overload failure that point toward the initiation of the cracking. The photo to the right shows chevron marks clearly indicating the fracture origin on the side of a chainfall hook.

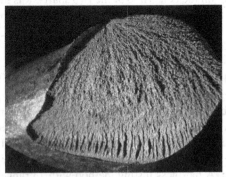

Brittle material: A material that, when stressed, breaks into pieces with essentially no visible plastic deformation. Typical examples are glass, ceramics, gray cast iron, and hardened steel.

Ductile material: A solid material that, when stressed, will have substantial visible plastic deformation before fracturing. Some common examples are silver, mild steel, copper, wrought aluminum and lead.

Elastic deformation: Distortion of a component caused by a force that, when the stress (force) is removed, will result in the piece returning to its original shape.

Failure: A component or a system that does not function at the design rate for an acceptable length of time.

Fatigue failure: A fracture that results from repeated stress (force, load) applications where the crack propagates across the part during those multiple force applications. Some common approximate groupings are:

GEAR TERMINOLOGY

Active profile: The side of the tooth contacting the teeth of the opposing gear (DriveR or DriveN) teeth. (Reversing gears will have two active profiles.)

Addendum: The portion of a gear tooth above the pitch circle

Backlash: If one of a pair of gears is locked in position, the backlash is the clearance between the driving and the next driven tooth. Some backlash is generally needed for lubrication to enter and escape the mesh, but there are very high precision "zero backlash" gears.

Bull gear: The driven (DriveN) gear, sometimes referred to as "the gear".

Dedendum: The portion of a gear tooth below the pitch circle. It is larger than the addendum by the depth of the root clearance.

Pinion: The driving (DriveR) gear, usually, but not always, the smaller gear.

Pitch circle: (also called the pitch diameter) – The ideal mating centerline of the gear.

Pressure angle: Referenced to the centerlines of shafts of mating gears, it is the angle at which the driving force is applied, i.e., the force that causes the rotation of the driven gear is applied at an angle of $14\frac{1}{2}°$ to the centerline. With early North American gears, the common pressure angle was $14\frac{1}{2}°$, today a common pressure angle is $20°$ and $25°$ gears are also common.

Root clearance: This is the ideal separation between gears when a tooth from the other gear in on the centerline of the two shafts. It is measured between the top land of the mating gear tooth and the body of the pinion (or vice versa).

HARDNESS TESTING TERMS

Brinell hardness: Hardness testing method that uses a 10 mm (~3/8") diameter ball pressed into a material. Typical indentation loads are 3000 kg and 500 kg. Value is abbreviated as HBW. (Older readings were abbreviated HBN when using a steel ball and HBW when using a tungsten carbide ball, but standards now call for the exclusive use of the tungsten carbide ball and the designation of HBW. It was also common to see the value abbreviated BHN.)

Rockwell hardness: A hardness testing method with 13 ranges that uses indenters that are lightly, and then heavily pressed into a part. Some of the ranges use ball-shaped indenters while others use diamonds and the values are abbreviated as HRA, HRB, and so on.

Vickers hardness: This hardness testing procedure is usually performed in a laboratory setting, because of the need for precision control, and has an extremely wide range of applications. In it, a pyramid-shaped penetrator is pressed into the piece with a variety of forces. The depth of penetration is closely measured and reported as a HV value.

HEAT TREATING TERMINOLOGY

Surface (case) hardened: The outer portion of the part is hardened while the central core is left much softer. Typically, the depth of the hardened area is less than 3 mm (~1/8") but some large gears may have a case as thick as 0.20" (5mm). Used almost exclusively for wear resistance. Several processes with carburizing the most common for industrial machinery.

Through hardened: The part is essentially the same hardness throughout. This is generally used for strengthening a material, but also confers additional wear resistance. It is usually less expensive than case hardening except on large parts.

Tempering: The process of reheating a previously heat-treated and hardened part to a lower temperature and holding it at that temperature for a time to effect metallurgical changes that both improve the toughness and reduce the internal stresses. Tempering is usually performed immediately after the initial hardening process to reduce the internal stresses.

Hertzian fatigue: The common design fatigue mechanism in gears and bearings with the cracks starting below and essentially parallel to the contact surface. (The German scientist Heinrich Hertz started the mathematics to describe how a fatigue crack can develop in material that appears to be stressed only in compression.)

Modulus of Elasticity: Also known as *Young's Modulus*, and is a measurement of a material's ability to deform with a given load. Within a close range, it is consistent within a given material family. For example, all steels have a modulus of elasticity of approximately 10^9 N/m^2 (29,000,000 psi).

Overload failure: A failure that results from a single extreme stress (force, load) application, i.e., the crack, or the distortion, grows across the part as a result of one force application. Failures normally fall into one of two categories:

Ductile overload failures pieces that show considerable plastic deformation upon failure.

Brittle overload failures pieces that display essentially no visible plastic deformation upon failure. The fracture face has a relatively uniform surface roughness. (Although the face of a fatigue failure will have the same fracture plane as a brittle overload, the fatigue failure will have a progressively rougher surface as the crack grows across the part.)

Plastic deformation: Distortion of a component caused by a force that, when the stress, i.e., the force, is removed, will result in the piece being permanently changed in shape by more than 0.2%.

Progression marks: Sometimes called *beachmarks* or *stopmarks,* these are irregular lines across the surface of a fatigue fracture face that indicate how the crack grew across the face, however they do not appear when the magnitude of the fatigue force is consistent. Their interpretation aids in understanding the forces that caused the crack. (*Progression marks* occur when the crack growth slows or temporarily stops and are a manifestation of the crack tip plastic zone frequently discussed in fracture mechanics.) The *progression marks* on the fractured gear in the photo to the right show the crack started near the left side of the gear tooth and for the heavy contact to be on one side, the gear had to be badly misaligned.

Ratchet marks: These are the boundaries between fractures that appear when adjacent fractures start in slightly different planes. As the fracture surface grows
the two planes eventually unite, but before that the *ratchet mark* serves as
divider between planes. The photo to the right shows a plate where the
cracking starts on the top surface. In the center of the photo are two *ratchet
marks* pointing in opposite directions. The initial crack was between them
and later *secondary ratchet marks* appear farther right and left.

River marks: These are marks that often appear on a fracture face, but usually well
away from the origin. They appear similar to what would be seen on a topo-
graphical map including the plot of a river and its tributaries and, similar to
the river's flow, *river marks* show how the crack progressed. The photo to
the right actually shows two sets of river marks. The ones at the top of the
picture are relatively coarse and not often seen while those in the lower left
are more typical. Both sets show a crack face that is growing from the top
to the bottom of the piece.

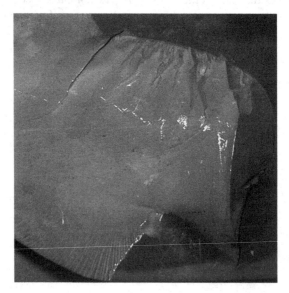

Stress: The load or force on a part divided by the area the load is applied to. If a part has a 14,000 pound force load and the area is 2 in^2, the stress is 7,000 psi (pounds per square inch). The common metric term is megapascal (MPa) where a pascal is 1 newton/square meter and a megapascal is 1,000,000 pascals. 100 psi = 0.689 MPa.

Stress concentration factor: Also known as *stress riser,* this is a change in shape or metallurgy that locally multiplies the stress in the part. Common examples include a step or a groove in a shaft and a hole in a plate.

Surface fatigue: Progressive deterioration and material loss initiated on the surface of a gear or bearing and resulting from repeated load applications. (This is damage to a gear or a bearing contact that begins on the surface and progresses inward, as opposed to *Hertzian fatigue* which begins subsurface.) Surface fatigue is a frequent result of poor lubrication and/or contact surface contamination.

Tensile strength (stress): The load at which a material will break into two pieces.

Yield strength (stress): The stress level at which a material will permanently deform by 0.2%.

Index

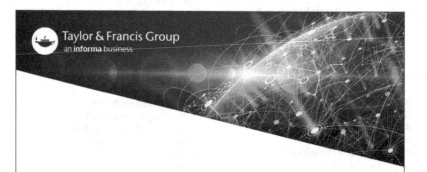

Printed in the United States
by Baker & Taylor Publisher Services